W9-AJO-632

LIAISONS of LIFE

LIAISONS of LIFE

•
•
•

From Hornworts to Hippos, How the Unassuming Microbe Has Driven Evolution

Tom Wakeford

John Wiley & Sons, Inc.
New York • Chichester • Weinheim • Brisbane • Singapore • Toronto

This book is printed on acid-free paper. ♾

Published by John Wiley & Sons, Inc.
Published simultaneously in Canada

Design and production by Navta Associates, Inc.

Library of Congress Cataloging-in-Publication Data:
Wakeford, Tom.
Liaisons of life : from hornworts to hippos, how the unassuming microbe has driven evolution / Tom Wakeford.
p. cm.
Includes bibliographical references (p.).
ISBN 0-471-39972-8 (cloth : alk. paper)
1. Symbiosis. 2. Evolution (Biology) 3. Symbiogenesis. I. Title.
QH548 .W25 2001
576.8—dc21 00-043911

Printed in the United States of America

10 9 8 7 6 5 4 3 2 1

Dedicated to my parents.

Contents

LIAISONS of LIFE

Introduction

As a child I took the natural world for granted. I lived in a village nestling in a valley checkered with dairy farms and woodland in the North of England. I spent my time climbing trees, paddling in streams, and picking berries on the fells. At eleven I cycled up the steep valley to the town for my secondary schooling. The thought of a long school day was lightened by a panorama of the Lake District peaks stretched out before me on the horizon. At sixteen I moved to Cambridge and the flat expanses of the East Anglian fens. The unfamiliar scenery made me yearn for escape routes. Fortunately I came across a group of a dozen young explorers who had organized a school party to Iceland, and they let me join them.

In this Nordic wasteland of fire and ice on the edge of the Arctic Circle, life was scarce, even in summer. Our base camp lay at the foot of Hekla, the volcano known to medieval Europeans as the gateway to Hell. From the tents we could see not a single tree, bush, or sign of animal life, only a bubbling spring of boiling-hot water. The toxic fumes from a nearby sulfur vent meant that we could observe its geological kaleidoscope of colors for only a few minutes at a time. Yet in the middle of this caustic cauldron we found thriving communities of purple, orange, and green bacteria. These bright microbial mats were layered on top of each other like a kind of lasagna. Intrigued by their survival skills and otherworldly beauty, I collected samples of them in a piece of newspaper and hoped they could endure the long journey back to my school biology lab. Dried, a bit mangled, but still robust, they did.

They faced a stiffer challenge in my favorite science teacher, Mr. Tomkins, who threw away half my specimens because he mistook them for dirt. The cleaning staff was another problem. Annoyed by the pungent smell of rotten eggs emitted by my Arctic guests, they spread a rumor that I was making explosives. Finally, in a converted fish tank, I managed to nurture my bacterial scum, and gradually the experience began to transform all my ideas about the living world. For under the microscope, the colored layers of microbes harbored an ecosystem as complex as anything I had seen.

Stephen Tomkins was also revealing hidden microbial dimensions to our practical classes. For years he had kept a fish tank at the back of the classroom—bone-dry all winter, with dried-up crud at the bottom. Each year he would fill it with tap water just once in the early summer. We watched spellbound as scores of tiny brine shrimp, whose microscopic eggs had lain desiccated and dormant for months, hatched and grew within a few weeks. Meanwhile the summer sun beat down and their water supply rapidly evaporated. Miraculously they reached maturity and laid a fresh batch of eggs just before the water vanished.

But what were they eating? Some of my classmates suspected Mr. Tomkins of secretly adding fish food to their tank. Instead he showed us the tiny microbes whose minute spores dried onto the surface of the shrimps' eggs. Once he added the water, he told us, these microbes began to multiply far faster than the shrimps, using light energy from the sun. The invisible bugs thus supplied the shrimps with a perpetual picnic. Even more remarkable, as I remember, our evolution teacher, Dr. Reiss, told us about a "new" theory that all our cells were the evolutionary product of intimate associations between bacteria billions of years ago. I was hooked.

When I went to university, however, microbes—the collective term for bacteria, fungi, and a disparate assortment of single-celled organisms called protists—seemed not to be on anyone's agenda. Of course these bugs' biggest mistake in terms of public relations is to be born so very, very small: most bacteria are little more than one-thousandth of a millimeter long. If a bacterium were enlarged to the size of a cigar, half an inch across by four inches long, a human magnified to the same extent would be fifteen miles in height. A Gary Larson cartoon that is a favorite with microbiologists is the one where two bacteria are pointing at something at the edge of the picture. "That's not your brother," says one to the other, "that's just an air bubble." The joke hints at the frustration generations of school pupils and university students have felt at the elusive nature of these bugs.

Some university scientists did share my microbial perspective on life, but their work tended to be pushed into the cracks between the big subjects of the day. Fortunately, a handful of researchers were studying these atoms of evolution, and some of these tutors would prove inspirational.

While most of my fellow biology students tended to favor research projects focusing on the big, powerful, or colorful, tutors such as Janet Moore encouraged us to study the small and seemingly insignificant. My favorite of her tiny research beasts were the tardigrades. Photographs make them look like miniature teddy bears, yet they are only a fraction of the size of a pinhead. Distributed throughout the world, yet hardly noticed, they live in drops of water attached to mosses and lichens. It was Janet who directed me to the relevant people and places in my quest for the hidden microbial dimensions that lay behind our degree syllabus. One of the first people she mentioned was Sir Vincent Wigglesworth, a nonagenarian éminence grise of everything to do with insects. His *Principles of Insect Physiology,* first published in 1934, was the first book everyone turned to when they wanted to know what was going on inside a particular insect.

In 1952, Wigglesworth conducted an experiment that showed that newly hatched onion fly larvae were unable to survive on onion tissue if it had first been sterilized. Despite apparently having all the food they required, the insects starved and died. However, if some *Bacillus* bacteria were added to the onion, the insects developed normally. It took another generation before biologists followed Wigglesworth's lead and turned their attention to the microbes that lived inside insects.

Not a single lecture in my degree course referred to the implications of Wigglesworth's experiment. At the bottom of a huge reading list on mathematical modeling languished a reference to an article in the journal *Science* in which a distinguished U.S. scientist heralded microbes, not deoxyribonucleic acid (DNA), as the most innovative element in evolution. An evolutionary theorist working in another department reckoned that a new microbially informed perspective might help plug the remaining gaps in the theory of evolution. But we were not made aware of her work.

Intrigued by these disparate people and theories, I embarked on a quest to discover whether their ideas hung together, or had been ignored because they were believed to lead nowhere. I read all I could on the evolution and ecology of bacteria, plants, animals, and fungi. Perhaps most important, I took a year's course studying the history and philosophy of science.

Historians suggest that a key factor preventing scientists from exposing these associations between plants, animals, fungi, and microbes was the lack of the appropriate laboratory technology to carry out critical experiments. "Good scientists," said Nobel prize–winning immunologist Peter Medawar, "study the most important problems they think they can solve. . . . It is after all," he concluded, "their professional business to solve problems, not merely grapple with them. If politics is the art of the possible, research is surely the art of the soluble. Both are immensely practical-minded affairs." Examining how microbes form liaisons with animals or plants is a quantum leap in experimental complexity from looking at how members of the two groups of large organisms associate with each other.

Another misfortune to befall microbes was the grim circumstance in which they received notoriety. Fame was thrust upon them during research not into flavors of cheese or the bouquet of wines, in which they are the crucial ingredients, nor in agriculture, where they ensure the fertility of crops. Instead they were immortalized by medical research, where their presence in the blood was seen as a sure sign of disease. Finding nonpathological microbes rather complex and confusing, the nineteenth-century French pioneers of microbiology, such as Louis

Pasteur, followed the path of least resistance, and greatest public fear, by studying those that appeared to be deadly pathogens.

Pasteur single-handedly spawned the antibacterial age. His cult of cleanliness prevented us from seeing our microbial associates as anything other than invisible assassins. Though some of his first and most notable achievements were to improve management of the highly productive bugs that were the active ingredient of the sugar distilling and wine industries, Pasteur is most famous for renaming bacteria "germs." Becoming increasingly shrill as his fame increased, he advised that these enemies of the people should be tracked down and destroyed.

Pasteur was undoubtedly a pioneer in tackling the bacteria that cause disease. In 1881 he devised a vaccine for anthrax, a fatal disease that threatened French sheep farmers. In 1885 he claimed to have invented a vaccine for rabies, though historians have recently questioned his account. His innovative techniques quickly led to the development of vaccines for tuberculosis (TB), cholera, and the plague. His pasteurization process for cleansing milk of TB bacteria continues to this day. Yet Pasteur's championing of the view that disease was solely due to bacteria and bore little relation to the malnutrition of the victim was controversial. Many doctors of the time argued that it was far more important to improve the nutrition and living conditions of the poor than to identify the bacteria that finally infected and killed them.

Pasteur would remain undaunted. An intensely political animal, Pasteur once stood as a candidate for the French Senate and he brought the instincts of the campaign to his war on germs. A conservative, he feared the mob—that faceless mass of peasantry whom Pasteur charged with murdering his king during the Revolution and instigating the Terror against the aristocracy. Writer and historian David Bodanis has demonstrated how the language Pasteur used against these masses was almost exactly the same as that which he developed to characterize bacteria:

> Let the mob take Paris and without the King or Emperor to shore us up we would dissolve into aimless bodies no different from the mob; let the bacterial mob take our physical body and we would decay into a putrefying bacterial mass no different from the attackers. If unpleasant entities such as the people or bacteria had to exist, then they must be kept firmly in their place. The people, especially the workers, were safe only if kept in passive Catholic trade unions, or state-run clubs, or other trustworthy bureaucratic bonds. The bacteria, in all their unpleasant and quick-to-grow varieties, were safe only if restricted to one slot in the Great Chain of Being, that of the decomposer of dead bodies, destroying order only after all life in it had naturally gone, and returning its atoms to the soil for re-birth. Outside of that, though, and they were terrible.

By the late 1890s the imagery of a loathsome bacterial mob had become such a part of everyday life that it became one of the most common metaphors used by early British war correspondents. As General Kitchener's army descended on Egypt, the *Daily Mail*'s reporter described the enemy as "swarming up the hill . . . like a torrent of death." During World War I, sections of the British press began to called the Germans "GermHuns," who, said the *New Statesman,* must be stopped as if they were "an epidemic of scarlet fever." The idea of surrounding an enemy army so completely that none of them can escape is still referred to with the bacterially inspired French phrase *cordon sanitaire.*

Pasteur's students testified how he could sow, cultivate, and domesticate his microbes so that the invisible became tangible to the ordinary onlooker. He was alone in being able to pacify the enemy within. Gerald Geison, a historian at Princeton University, calls him the "artist of the invisible world." But Pasteur's legacy of scientific genius, his status as a would-be microbe killer, and the biological words he coined in support of his theories would cast a shadow over microbiology for the next seventy years.

Epidemics of disease afflicted those successive waves of European immigrants who, often malnourished, arrived in the United States, first from Ireland, Germany, and Scandinavia in the nineteenth century, then from Italy, Eastern Europe, and elsewhere continuing well into the twentieth century. Little surprise, then, that among the most enthusiastic followers of Pasteur's germ theory of disease were doctors in the cities of the U.S. East Coast. In the 1920s and 1930s this enthusiasm reached fever pitch. At the same time as Stalin was purging his country of potential intellectual "enemies within," America suffered a collective wave of antimicrobial paranoia, fed by the newspapers and books of the day. Most famous of these was *The Microbe Hunters,* by Paul de Kruif, published in 1926.

De Kruif's book is a brilliant crime novel, in which scientists are cast as the defenders of private property, while bacteria play the part of thieves and murderers. "These wretched microbes," he wrote, "kill millions of human beings mysteriously and silently." He concluded that these germs were "more efficient murderers than the guillotine or the cannon of Waterloo." His words influenced successive generations of industrial civilization, yet we now know that disease-causing bugs make up only a tiny fraction of the total diversity of microbes.

A randomly chosen bacterium is far more likely to be digesting your domestic sewage or supplying oxygen for you to breathe than it is to be making you ill. The vast majority of bacteria live harmlessly and unobtrusively on the surface of our seas and in our gardens, quietly undertaking biological processes crucial to every ecosystem. Many species are essential to the production and digestion of almost all of the kinds of food that we eat, the recycling of our wastes, and the fertility of our soils.

Some bacteria do kill, though ironically they often do so by producing toxins, which are a sign that they themselves are under stress. Metaphors that cast all bacteria as germs have obscured our biological understanding for too long. In reality, pathogenic microbes are the exception, not the rule.

By the time I took my final exams, the world of conventional "big biology" seemed a world away from all this. Virtually everything seemed to be turned on its head by taking a microbe-eyed view of life. Microbes are not the deadly killers that must be cleaned from every nook and cranny at the first opportunity, but have rather been crucial innovators in the past 4 billion years of evolution. It came as no surprise to learn that the eminent Harvard University ant watcher and biologist E. O. Wilson, when asked what he would do if he were to have his time again, replied that he would like to have been a microbial ecologist. Because of the minute scale and speed of their activities, the bug's life—spent on our teeth, in our guts, or in our oceans—is only just beginning to be uncovered.

The diversity of microbes is even greater than that of insects. Vigdis Torsvik, a bacteriologist at the University of Oslo, Norway, has found that an average teaspoonful of soil contains around ten thousand different genetic types of bacteria. If you extracted all the bacteria from two acres of farmland, their total weight would be greater than that of one hundred sheep. Another area in which microbes may outdo the animal kingdom is in their extraordinary sex lives.

Unlike other aspects of their lifestyle, the sexual antics of bacteria have been well researched over the past fifty years, mainly because of their relevance to treating human disease. If it were not for sex, the life of a bacterium would seem rather uniform. Animal or plant cells can change into a million different cell types—a petal, an eye, or a skin cell. In contrast, a bacterium stays the same size, and the same shape, almost all its life. While any given mammal cell is likely to draw on only 5 percent of its genes during its life span, a bacterium uses most of its genes, most of the time. Because they reproduce so frequently and at such speed, bacteria that carry too many genes waste energy copying them every time they reproduce. They would be quickly overtaken by other less encumbered strains. Microbial geneticists have found that bacteria undergo a kind of gene-injecting sex that occurs quite independently of the bacterium's reproductive cycle, allowing them to import new genes throughout their lifetime without waiting to reproduce.

Popular accounts of the natural world have found the bacterial scale of life difficult to depict. Yet by retreating into a mere focus on fierce or fuzzy creatures they risk distorting our understanding of ourselves and our place in biology. By concentrating on mammals, they reinforce the suspect notion of evolution as progress, inexorably moving toward its summit—humanity. The suggestion that bacteria are primitive beings that long ago passed the torch of evolutionary innovation on to large organisms is misguided. Bacteria are the eternal innovators in the history of life. Biology without bacteria is as incomplete as physics without forces.

As our continual medical battle with disease testifies, bacteria have shown themselves to be our equals. They not only invented most biological processes—everything from sex and light reception to respiration and movement—but almost every biologically generated chemical on Earth. Bacteria also spearheaded the evolution of the other four kingdoms of life. Every time we put microbes to use, whether in making yogurt or cleaning up oil slicks, we are drawing on 4 billion years of incomparably sophisticated yet subvisible innovation. It is no surprise that out of the three diseases virtually defeated by modern medicine—smallpox, polio, and measles—none are microbial (they are viral), whereas of the three biggest current killers—malaria, TB, and AIDS—two are.

Microbes and their alliances are fundamental to the origin, evolution, and current function of every creature we encounter, from the hornwort to the hippo. Even our understanding of ourselves is greatly advanced by taking a new microbe-eyed view. Today it makes no sense to study evolution or ecosystems, be it in our garden soil or at the bottom of the Atlantic Ocean, without recognizing the keystone activities of our microscopic cousins. Indeed, three of the most important breakthroughs of the past half century—symbiogenesis theory, microbe-mediated immunity, and the Gaia hypothesis—have all challenged traditional biological theories by uncovering the hidden powers of the microbial realm.

Our newfound knowledge of microbes, and the intimate liaisons in which they become involved, is fueling a transformation in the scientific worldview. The chapters that follow chart the development of an integrative history of life on Earth from this new perspective, all the way to Hollywood. Amid scenes of pod racing and intergalactic politicking, *Star Wars: The Phantom Menace* reveals that the Jedi's mysterious Force arises from the symbiotic relationship with "midichlorians," microscopic life-forms that reside (according to George Lucas) "in all living cells."

In biology, symbiosis is the term used to describe long-term intimate associations between different organisms, usually involving microbes. Such alliances are fundamental to the development of every living system. Drawing on new evidence from creatures found in settings as diverse as underwater volcanoes, termite mounds, and the gaps between our teeth, this book argues that staying alive is as much about bonding with your neighbors as it is about growing and reproducing.

Despite having been first proposed more than a hundred years ago, the idea that liaisons with microbes are a primary means of evolutionary innovation has taken most of the intervening years to reach center stage. Each chapter explores the mixture of personality politics, technological backwardness, and blind ignorance that caused the revolutionary ideas of a handful of pioneers to be condemned as heresies, but which are celebrated today as some of the greatest breakthroughs in the history of science.

In 1896, Beatrix Potter came face-to-face with the first barrier for the symbiosis pioneers—ignorant prejudice. The soon-to-be-famous children's author and illustrator was hounded out of biology by the closed ranks and narrow minds of London's top scientific institutes. Their members, all male, refused to accept Potter's evidence that lichens, those curious encrustations living on tree trunks, seashores, and walls, were made up of not one but two organisms in intimate alliance. Yet her insight was more far-reaching than either she or her contemporaries could ever have dreamed.

Not only lichens, but also almost every tree, bush, and grass on Earth leads a double life—married to a fungus. While orchids and oak trees appear to be individuals, in reality they live in an inextricably interwoven liaison with a worldwide web of underground fungal foragers. A century after the stifling of Beatrix Potter's search for natural truths and her final retreat into voluntary exile, we can now celebrate her radically new view of life. Interconnectivity can be a strength rather than merely a source of potential conflict; a vital resource rather than a drain.

Even after the great man's death, Pasteur's influence pervaded not only the medical laboratories of Europe and North America, but even the study of the evolution of marine life. Many of the most spectacular examples of animal symbioses are buried deep in the ocean abyss, but much of the work of the early investigators into this astonishing realm was crushed by Pasteur's supporters. In the weird world of the seabed, biologists have now discovered a spectrum of luminous animals. New genetic techniques have revealed that their glow is a gift from their bacterial associates. While deep-sea fish have their own microbial searchlights for hunting in the blackness, octopuses and squid use these glowing symbionts as a disguise from predators, or to chat up the opposite sex.

In the case of insects, the latest molecular tools, such as DNA fingerprinting and luminescent gene markers, have exposed the pervasive influence of bacteria on beetles, butterflies, and blowflies. They have power over wide-ranging aspects of an insect's life, including sex determination, digestion, and nest construction. Genetic analyses have also revealed the extraordinary gardening exploits of the ants. By building their own self-fertilized fungal farms, cultivating a continual harvest, and swapping favorite crop varieties with their neighbors, ants have achieved a sophistication of microbial alliance that may have lessons for the way we feed ourselves. Some researchers predict that one day symbiotic bacteria may even be shown to provide the mechanism behind the extraordinary diversity of insect life on Earth.

Historians of science have puzzled over why these fundamental insights took so long to receive recognition. Primitive analytic tools, Pasteurian paranoia, and intellectual inertia are part of the explanation. But, tragically, the study of symbiosis also fell foul of global politics: world wars, nationalism, and anticommunism, to name a few. Symbiosis was invented as a purely scientific term, but it was fatally bracketed in the minds of its enemies with dangerous political movements. Nor did it help that the pioneers in this field were largely from non-English-speaking countries such as France, Germany, and prerevolutionary Russia. In the wake of the carnage of World War I and the new threat from the Soviet Union, symbiosis was condemned by mainstream science as a political subversion that could provide explanations neither for humanity's apparent lust for conflict nor for the evolutionary patterns of life. Symbiosis became an international pariah subject, the victim of tacit textbook censorship and McCarthy-like witch-hunts among professional scientists.

In the 1990s, symbiology at last escaped across biology's own Berlin Wall. Those who had championed the significance of symbiosis for decades finally saw their ideas triumph, not just because new genetic techniques strongly supported their theories, but also because their fellow scientists became more open to a symbiotic perspective. The latest research poses exciting challenges to received scientific wisdom. Far from all organisms being in constant competition, many seem to have bonded to such an extent that it is no longer possible to tell where one ends and another begins. In changeable environments such as the forest floor, it seems that microbes provide resources to their extended family according to the recipient's needs.

The sudden transformation in symbiotic understanding during the past few years heralds the final hurdle for life's liaisons: the challenge they pose to traditional evolutionary theory. Brought up on the orthodox story of life evolving via chance mutations and competitive struggles, biologists struggled to comprehend a world of microbial mergers and emergent wholes. Newfound knowledge of widespread intimate association in nature is now adding a new dialect to the language of Darwinism.

A model of life that recognizes the key role of our associates offers a new scientific tool kit for the twenty-first century. It helps us resolve the tensions between the holistic and reductionist views of life, the role of nature and nurture in development, and the relative importance of the individual and its community.

Both in biological science and in our everyday life, we have the beginnings of a new relationship with our planet based on a more complete understanding of wild and domesticated nature. This new perspective recognizes the centrality of microbes in all living systems, our inescapable connectivity with them, and the importance of maintaining liaisons with our neighbors and associates. Having reclaimed a radically new way of looking at life after a century during which it was kept at the margins of science, *Liaisons of Life* suggests how we can build new alliances for the future.

CHAPTER 1

BEATRIX versus the BOTANISTS

Where observation is concerned, chance favors only the prepared mind.

LOUIS PASTEUR, 1854

Had Beatrix Potter been allowed to follow her vocation, Peter Rabbit and Mrs. Tiggy-Winkle might never have been born. Instead of writing and illustrating stories loved by children all over the globe, she would have been writing groundbreaking articles for scientific journals. Beatrix's ambitions were thwarted not only because she was a young woman attempting to contribute to a profession almost entirely dominated by Victorian men, but also because she was a symbiologist—a proponent of the dissident theory that some organisms were composed of not one but two different beings. Her story has become a legend of youthful scientific inquiry stifled by pomposity and prejudice, and of a heresy that was later vindicated.

No one who has encountered Beatrix Potter's graphic contributions to biology is surprised at the beautiful natural detail of the pictures in her books, such as *The Tale of Squirrel Nutkin* and *The Tale of Jemima Puddle-Duck.* They depict animals, dressed as human beings, immersed in the fairy-tale activities of a sanitized Victorian village. Virtually every picture includes features of the surrounding landscape—a carpet of wildflowers, the rough bark of an oak tree, or a mossy stone. During her own childhood, she loved nothing more than recording the subtle details of the living world. Often neglected by her parents, who like many Victorian aristocrats had decided not to send their daughter to school, she used to take herself on expeditions to the Natural History Museum, just round the corner from their London house. Here she would sketch anything that caught her eye, returning to the museum day after day. Animals were to form the central characters of her books, but as a child she would be just as enthralled by the fine structure of the gills on the underside of a mushroom as by the arrangement of feathers on a bird's wing.

Beatrix's childhood holidays were spent in Perthshire, Scotland, and in the English Lake District, where she was in her element. Here, accompanied by her brother, she could explore the natural world and draw exactly what she wanted. She soon developed into a superb watercolor painter, able to produce pictures that were both aesthetically satisfying and scientifically accurate. As she gained confidence, Beatrix started to compare her findings with those of other botanists of her day. In Perth she befriended the local postman and amateur naturalist Charlie Macintosh, whose studies of mosses and fungi can still be seen at Perth Museum. Macintosh liked her drawings. "His judgement," wrote Beatrix in secret code in her diary, "speaking to their accuracy in minute botanical points, gave me infinitely more pleasure than that of critics who assume more, and know less than poor Charlie. He is the perfect dragon of erudition, and no gardener's Latin either."

Middle-class Victorian society was in awe of the apparent power and moral superiority of the scientific worldview. Keen to make her own contribution, Beatrix kept careful notes of everything she saw, and compared them with what other naturalists had observed. During her teens, she also made detailed studies with her microscope of various botanical specimens. She quickly discovered that one of her favorite subjects of study, the lichen, was the battleground of an increasingly heated scientific controversy.

Lichens are the crusty green and gray covering of rocks and tree trunks. Often confused with the damp-loving mosses, lichens are less intricate but usually longer lived than their larger cousins. Worldwide, these lowly lifeforms cover ten times as much of the earth's surface as tropical rain forests. From the Arctic to the Tropics, lichens are a biological realm hidden in plain sight. Thousands of different species range in color from orange and black to brown and green. On a walk along a rocky coastline you might see only a handful of different wildflowers, but you could have walked past 80 or 90 different lichen species. With their variety of rock types, shading, and nutrients, British churchyards are paradises for lichens, with 180 different kinds having been counted around a single church.

Carolus Linnaeus, the eighteenth-century botanist and the founder of modern plant taxonomy, had no time for lichens. He called them the "poor peasants of the plant world." Thinking them to be either a primitive moss or an unusual fungus, his successors dismissively classified these curiosities as the "lower plants." The Swiss botanist Simon Schwendener begged to differ. In 1869 he startled the scientific world with a "dual hypothesis" for the taxonomy of lichens. He proposed that all members of the group came into being via the liaison between a fungus and an alga. The alga was useful to the fungus, which nurtured it, he claimed.

Most biologists treated Schwendener's ideas with contempt. They could not believe that even the most bizarre form of parasitic relationship could lead to a permanent merger between two organisms. It also seemed extraordinary to suggest that a composite organism could function as an integrated and successful whole. In the introduction to his classic *Lichen Flora of Great Britain, Ireland and the Channel Islands,* the Reverend Leighton scoffed, "I have purposely omitted any mention of the Schwendenerian Theory of Lichens, as I cannot but regard it as purely imaginary, the baseless fabric of a vision."

All the studies she made of the fine details of lichens, algae, and fungi drew Beatrix to share Schwendener's conclusion: lichens were made up of two completely different kinds of organism. Yet opposition to the dual hypothesis by the British Empire's most eminent botanists was becoming fiercer than ever. Finding it inconceivable that organisms could somehow be a mixture of creatures from different kingdoms, they clung to the idea that all organisms were exclusively either animals or plants. Reverend James Crombie, a prominent English naturalist, scolded: "A useful and invigorating parasitism—who ever before heard of such a thing?" So vocal were these scientists against what Crombie dubbed "this sensational Romance of Lichenology," the "unnatural union between a captive algal damsel and a tyrant fungal master," that they even influenced everyday speech in the late Victorian era. Calling somebody a "Schwendenerist" became a term of abuse for someone who dithered between two competing explanations for an event. "Even if endorsed by the nineteenth century," remarked botanist M. C. Cooke in 1879, such ludicrous symbiotic ideas "will certainly be forgotten in the twentieth."

At first, Beatrix was unperturbed by this opposition. Her uncle, the chemist Sir Henry Roscoe, had confidence in her and her belief in Schwendenerism. He urged that she give a paper at a scientific society, such as the Linnaean. Housed behind a grand facade in London's West End, the Linnaean Society was an international forum for naturalists and evolutionary biologists, as it had been when Charles Darwin and Alfred Russel Wallace had announced their theories of evolution there earlier in the century. But while Roscoe could contribute to the society's proceedings, Beatrix was, like all women, barred. Although now in her late twenties and gaining a reputation for her acute observations of nature, she was not even allowed to attend the society's open meetings. Her uncle eventually won the right to read her paper at a meeting of the society himself, but the official record of the meeting has been lost. We can only imagine the mixture of smirks and tut-tuts that greeted her findings. Beatrix, already a shy and reclusive character, recorded in her diary her feelings of humiliation at her treatment. Worse was to follow.

In 1896, just after the Linnaean debacle, Beatrix made an appointment with Mr. W. H. Thiselton-Dyer, director of the Royal Botanic Gardens, Kew, in order to show her drawings to him and his staff. Many were vivid illustrations supporting Schwendener's radical lichen theory. Although she expected the Botanic Gardens' staff to be skeptical of her observations, the sexual and scientific prejudice she found at Kew shocked and upset her. Having seen the outdated and demeaning uniform that female staff were forced to wear on her many previous visits to the gardens, she already had an idea of what treatment she was letting herself in for from the director. "I fancy he may be something of a misogynist," she wrote, "*vide* the girls in the garden who are obliged to wear knickerbockers."

When Beatrix entered his room, Thiselton-Dyer was puffing on his cigarette. He ignored her and started boasting to her uncle about the gardens' hyacinths being even better than those in Holland. When Beatrix pressed him for an opinion on her drawings he refused to even look at them, referring her instead to the botanical gardens at Cambridge. "I fear he is jealous of outsiders," wrote Beatrix afterward, "But it is odious to a shy person to be snubbed as conceited, especially when the shy person happened to be right, and under the temptation of sauciness."

After what she describes as a "storm in a teakettle" she left Kew, never to enter the world of professional biology again. She disliked what she sarcastically called the "grown-up world" of science. Two venerated institutions, which had embraced the theory of evolution by natural selection forty years earlier, were now shattering the aspirations of one of Darwin's most able successors.

Beatrix was dispirited that the excitement of the last few years, and the great hopes of being able to make a significant contribution to science, had so cruelly and abruptly been snuffed out. She knew her passion for lichens and fungi would now only lead to further public ridicule, both for herself and for her favorite uncle. One by one, she laid her treasured folios of watercolors aside. It would be 1967 before William Findlay, president of the British Mycology Society, returned them to their rightful place as outstanding scientific studies of nature, when he used them to illustrate his field guide to the fungi and lichens of the British Isles.

Despite their initially vicious reception, the ideas of Schwendener and Potter were accepted by most biologists within a few decades. Lichens were shown to be true dual entities, the association of a bacterium or an alga with a fungus. In 1929, H. G. Wells and Julian Huxley remarked in *The Science of Life* that "a lichen is no more a single organism than a dairy farm is a single organism." In this encyclopedic textbook, the authors describe a diverse range of similar alliances that had been recorded in a wide variety of plants, animals, and fungi. Far from being primitive taxonomic obscurities, irrelevant to the rest of evolution, lichens could be, they suggested, the dual ancestor common to all plants. More recently, Schwendenerism has not only been rehabilitated, but has provided the key to understanding the role of intimate associations in the evolution of our plant-dominated landscapes. Arriving with their radical networking manifesto 400 million years ago, fungi are the alliance-building kingdom that built the power supply for almost all terrestrial life.

In the world of picture books, lichen illustrations on their own had little chance of being appreciated. But in her 1904 *The Tale of Benjamin Bunny* and many subsequent books, Beatrix Potter included trees in the background to her main story whose bark was studded with luxurious encrustations. In editions printed after her death, however, this last legacy of Beatrix's earlier intellectual bravery was erased. Now, as these kinds of alliances between organisms begin to be celebrated rather than dismissed, it is surely time for the original lichen-enriched editions to be returned to nurseries and playrooms around the globe. Having received posthumously an official apology from the Linnaean Society for its treatment of her, at a meeting held in her honor in 1997, exactly one hundred years after it had barred her from speaking, Beatrix Potter is now beginning to receive the recognition she so richly deserves.

Early Schwendenerists would have been emboldened in their convictions if they had had the benefit of current insights into the chemical logic and hand-in-hand evolutionary consequences that underlie the lichen's liaison. On one hand there are the algae, the common name for yellow, brown, red, and green microbes. They manufacture themselves almost out of thin air—from water, carbon dioxide in the atmosphere, and the absorbed rays from the sun. The ability to make their own food autonomously by the process of photosynthesis has led scientists to describe them as autotrophic—literally "self-feeding"—a category in which they are they are joined by many kinds of bacteria. On the other hand are the fungi, which along with most animals depend, by contrast, on various kinds of "heterotrophic" metabolism. In other words, mushrooms, like mice and monarch butterflies, derive their nutrition from compounds made by other organisms.

Intimate liaisons often arise between organisms of the auto- and hetero- classes of trophism. Such relationships are commonly known as symbioses—literally meaning "together-living"—the long-term association of two different organisms. Why does this happen? One of the key reasons is that, at the microchemical level, the cells of all organisms slowly leak their contents like tiny tea bags. In lichens, green algae cannot prevent some sugars from passing through their cell walls, while fungi leak essential nutrients such as nitrates and phosphates through their outer membrane. Because most chemicals leaked by one associate are useful to the other, the integration of the two over evolutionary time has turned the passive leakage into a process of active transport by which means materials are passed at a far higher rate than natural diffusion, and both algae and fungi are provided with ready food generation after generation.

In addition, the lichens' particular lifestyle allows them to tolerate a wide range of conditions of intense cold, heat, light, and barren rock, which neither the fungi nor the algae on their own, let alone larger plants, could tolerate. When Edmund Hillary and Tenzing Norgay reached the summit of Mount Everest in 1953, they might have been surprised to notice that lichens, too, had scaled its heights. Often visible only as black specks that you would be forgiven for mistaking for pieces of soot, they can be found on almost any piece of rock on Earth.

In the year 1066, a more noticeable lichen served as an important landmark in Britain. As King Harold tried in vain to fend off the Norman invasion at the Battle of Hastings, he instructed his noblemen to assemble with their armies at the "hoar" apple tree on Caldbec Hill, meaning a tree that was gray and shaggy with lichen. The British still use expressions such as hoarfrost, meaning a frost that covers grass and tree twigs with a fur of ice crystals.

Lichens are a powerful symbol of endurance as well. In one of his sonnets, Shakespeare identifies lichens as the organism that betrays the age of rocks:

> Not marble, nor the gilded monuments
> Of princes, shall outlive this powerful rhyme;
> But you shall shine more bright in these contents
> Than unswept stone, besmear'd with sluttish time.

And an unknown medieval elegist meditating among the Roman ruins in the city of Bath describes lichens as a model of stoicism in the face of changing fortune and bloody wars:

> Time and again this wall endured,
> Lichen grey and red stained,
> As kingdom follows kingdom follows kingdom.

Across Renaissance Europe, knowledge of lichens contributed to a revival of interest in herbal medicine. John Gerard's *The Herball, or Generall Historie of Plantes,* published in 1597, contains woodcut pictures of five lichens, including tree lungwort and coral moss, along with their medicinal uses. According to the "Doctrine of the Signatures," which held that the Creator had marked those plants suitable for treating diseases by a resemblance to a specific part of the human body, lichens found growing on human skulls were considered a uniquely potent cure for epilepsy. Many more conventional lichen-derived medicines are still listed in the drugstore pharmacy manuals. Antibiotics such as usnic acid, found in the lichens *Evernia* and *Usnea* and over half all other lichen species, are used in products such as skin creams and antidandruff shampoos. Iceland moss is a lichen widely used to brew a tea that is effective against respiratory infections and to make throat lozenges. And as every science student knows, lichen extracts were the origin of litmus paper—the most simple means of testing whether a liquid is acid or alkaline.

During the height of the Napoleonic Wars, lichen dyes played a role in a famous military victory. In February 1797, a French army landed in Fishguard, Pembrokeshire. The British army was nowhere to be seen, yet the nervous French soldiers mistook the red-lichen-dyed cloaks of a number of distant Welsh women mounted on hill ponies for the uniforms of advancing battalions of regular soldiers. They surrendered without a shot being fired.

More recently this crusty clan, which together produce over four hundred compounds found nowhere else in nature, have been used to monitor pollution. The different lichen species present in a particular location act as a litmus paper for the atmosphere. Just a 5 percent increase in sulfurous oxides eliminates one leafy group—the foliose lichens. But like many organisms engaged in intimate association, the correct identification of different species demands a keen and sympathetic eye.

The same crustlike lifestyle that leaves lichens so easily ignored also allowed them to be the first significant colonizers of the land, perhaps as long as a billion years ago. Though bacteria were already present, living within cracks in rocks and at the edge of shallow pools, the lichen invasion began a more fundamental transformation in terrestrial ecology, that from rock to soil.

Today it is clear that the only scenario more remarkable than the evolution of the lichen would be one in which this joint organism had not evolved; the metabolisms of algae and fungi complement each other perfectly. Plugging into a range of ecological services via their intimate alliances, the plant kingdom has become the master of evolution by association. Neither Victorian prejudice nor the marginalizing of her favorite life-forms by twentieth-century evolutionists could prevent Beatrix's symbiotic manifesto from finally proving its scientific worth. Nature may, as Darwin seemed to imply, be red in tooth and claw, but it also, as Beatrix taught us, survives by being green-fingered.

In her last secret diary entries written soon after her rebuff at Kew, Beatrix Potter wished she could study the fungi and lichens again someday. "The funguses will come up again and the fossils will keep," she wrote. "I hope I may go back again when I am an old woman." By the age of thirty-nine she had. Potter bought a farm in her beloved Lake District, from where she spent forty years writing her children's books and enjoying the local farmers' tradition of breeding Herdwick sheep. Even if no more than a handful of the thousands who flock annually to Hill Top Farm understand that the real intellectual passion of her life was not Jemima Puddle-Duck and Benjamin Bunny, it is still comforting to know that this brave woman, so poorly treated as a scientist, was finally able to spend half her life there studying her forgotten symbionts away from the gaze of "grown-up" science.

CHAPTER 2

The WOOD WIDE WEB

> Ever since Darwin, the development of theory in ecology and evolution has been implicitly constructed for fruit flies, birds and people. . . . Organisms which commonly dominate much of the land and sea, and do not commonly display such determinate characteristics, have largely been ignored.
>
> LEO BUSS, *THE EVOLUTION OF INDIVIDUALITY*

In Ireland in 1845 a million peasants died and another million and a half fled the country, in the wake of a devastating famine. A potato blight had virtually destroyed the country's entire crop, yet the best researchers of the day could not find the cause. Some suspected soil-inhabiting fungi, while others believed that such microbes appeared only after the onset of the disease. Everyone feared the unknown culprit might strike again.

Insight into the disaster would only come several years later, with the publication in 1853 of *Understanding Plant Disease,* written by a twenty-two-year-old German biologist named Anton de Bary. His work took the sleepy world of botany by storm. Trained in rigorous experimental methods, de Bary undertook painstaking studies that laid the foundations of the new disciplines of plant pathology and mycology—the study of fungi. Moved by stories of the horrific effects of the famine that he had read during his childhood, and provided with his own botanical field station by his physician father, de Bary spent much of his teenage years studying plants and their disease. Now he showed step-by-step how fungal spores penetrated plant leaves, then reproduced inside their tissues, and eventually overwhelmed the plant's physiology, causing its untimely death.

After more than a decade of scientific commissions, official inquiries, and inconclusive research, neither Britain nor Ireland had been able to establish the role of the fungus. The Irish government and the Royal Agricultural Society of England begged de Bary to bring his by-now renowned expertise to the subject.

While most scientists at the time dismissed farmers' knowledge as uninformed and irrelevant, de Bary had long suspected as sound their centuries-old belief that a common plant called barberry harbored the blight. With his trademark thoroughness and open, yet critical mind he soon conducted an elegant series of experiments that enabled him to chart the whole life cycle of the offending microbe, which he named *Phytophthora infestans*—and which was indeed harbored by barberry plants.

De Bary went on to determine the mechanisms behind scores of economically devastating plant diseases such as wheat and rye rust, but it was not this work that most marks him as a remarkable biologist. De Bary's brilliance was that he went beyond both Darwin's exclusive focus on plants and animals, and Pasteur's preoccupation with microbes as germs that cause disease. He had the vision to see that the intimate associations of microbes with plants and animals were just as likely to lead to mutual dependence and innovation as they were to mutual destruction.

In 1878 de Bary gave a lecture to the Association of German Naturalists and Physicians that explained the meaning of a new scientific word—symbiosis. First coined a year earlier by a colleague, the German botanist Albert Frank, the term "symbiosis" was intended to embrace all the cases in which two different species live together, one living on or in the other. Rather than implying anything about the balance of costs and benefits between the associates, it would, according to Frank, be "based on their mere coexistence."

The original concept did not imply any necessary cooperation between organisms, any more than it implied exploitation. Symbiosis merely involved the intimate physical association of one creature with a different organism. Each symbiont influences the growth of the other, to which it is physically connected for much, or all, of its life. Many of the evolutionists skeptical of the significance of symbiosis assume that for such an outcome to be common, its evolution would have to involve some kind of collective foresight by the organisms involved. Yet, on the contrary, such liaisons persist by the same process of natural selection as all other evolutionary events.

Within months of the publication of de Bary's lecture, biologists all over Europe began to realize that phenomena they had put aside as evolutionary quirks unworthy of further investigation could be instances of what de Bary had described. Lichens could thus be put under the same conceptual umbrella as associations previously believed to be cases of parasitism, such as potato blight. A new coherence had been brought to a realm of science that had seemed disparate and confusing. Researchers beat a path to the door of de Bary's laboratory at the University of Strasbourg in the hope of learning more of this new symbiotic perspective.

Today we know fungi are most often the essential allies of plants, especially in the hidden realm of roots. Roots may not look inspiring, yet they are the biological power brokers that underlie the terrestrial triumph of the plant kingdom. By forming alliances with other soil organisms, roots have almost single-handedly given green plants dominance over the land surface of the earth.

When it comes to making the most of root-fungus alliances, orchids are perhaps the champions. Orchid seeds are so small that a million of them weigh just two grams.

They contain nothing apart from an embryo and an aspiration to meet the right mold. For without the participation of their own dedicated species of fungal associate, orchid seeds cannot even successfully germinate in the wild.

Even when fully mature, the orchid's root system appears pathetically small. Yet its fungal symbiont forms a large and dynamic foraging web that ensures the orchid's nutritional needs are fully met. In turn, it may receive small amounts of vitamins and nitrogen compounds from the plant. The orchid's generosity has well-defined limits, however. The plant keeps the fungus in check with natural fungicides, should it show signs of attempting to stray upward from its normal home inside the roots, to colonize the orchid's stem.

Conserving wild orchids presents special challenges, with at least a third of Britain's fifty species of wild orchid under threat. One of the most beautiful native species, the lady's slipper, is down to a single plant in the wild. But if conservationists are to succeed in growing rare specimens in the laboratory and greenhouse for subsequent reintroductions to the wild, it is up to conservation scientists to find the fungal partner for each endangered species. This is no easy task. Orchids are easy to identify, especially once they flower, but their symbiotic fungi are in a different league altogether.

English Nature, one of the British government's statutory conservation bodies, has now set itself the task of safeguarding as many of the remaining species as it can. Grace Prendergast, with Peter Roberts and Reem Salman at the Royal Botanic Gardens, Kew, has the unenviable task of finding which of hundreds of species of fungi, many of which do not even have names, will form a symbiosis with one of the endangered British orchids. Prendergast's team tests orchid-fungi compatibility by putting a few of the pinprick-size seeds of a particular species in the middle of a small dish of sterilized nutrient jelly. On the edge of the dish they place a small colony of a species of fungus. The seeds soon prepare to germinate, but with no food source are unable to grow. But from the edge of the dish a web of fungal threads, each far smaller than a fiber-optic strand, weaves its way toward the embryonic plant. Prendergast and her colleagues then cross their fingers. Will the two organisms be compatible? If they are, then when the thread reaches the embryo there is a biological equivalent of a flash of lightning, and both organisms start growing a hundredfold faster. A successful symbiosis has been established, and the orchid seedling can be grown into a mature plant, ready to be reintroduced to the wild.

To test each endangered orchid with every one of their collection of fungi is going to take years. In the meantime, Prendergast uses a synthetic cocktail, developed by Svante Malmgren, a Swedish pediatrician and amateur mycologist, which successfully mimics the chemicals that orchids normally milk from their molds.

It is a complex mixture of amino acids, potato stock, and pineapple juice. This method has allowed Prendergast to cultivate the lady's slipper orchid in the laboratory in sufficient numbers to be reintroduced to suitable habitats. However, only a handful have so far survived what she refers to as the "weaning" stage. Without the right fungus, no orchid is ever going to thrive.

Apart from habitat destruction, covert collectors are the biggest threat to endangered orchids in the tropics. The collectors are usually ignorant of the symbiosis on which the orchid depends for its survival. Though they smuggle the orchid to purchasers in Europe, the United States, or Japan, the fungal symbiont may not survive the journey. When the black-market purchaser tries to grow the orchid, any amount of fertilizer or watering may not revive it. The soil into which the orchid is planted is likely to be the wrong acidity for orchid growth and very unlikely to contain the compatible species of symbiotic fungi to replace the one that has died in transit. The whole illegal operation is often a pointless and destructive waste.

Once it would have seemed quite extraordinary that any plant could be so dependent on another organism that it cannot even begin life without it. But now it is clear that throughout their evolution, few plants have been without such an association.

From rotting leaves to insect droppings, the soil contains many nutrients, but they are always patchy in their distribution. With much of the soil barren, and nutrients existing only in unpredictable pockets, mature plants face a dilemma. Root construction is highly nutrient-demanding, and the chance of finding sufficient resources to justify growing more is usually small. The answer to the plant's dilemma is ingenious: to go online, connecting to the incredible fungal web that lurks in the soil.

The minute living threads of a fungus can extend hundreds of feet beyond its plant associates. Finding a source of nutrients, the fungus distributes them throughout its diffuse body. The plant absorbs its share through a fungus-root attachment that biologists know as mycorrhiza, named after the Greek for fungus and root.

Mycorrhizal networks give plants new ways of obtaining access to key nutrients that would otherwise be in short supply, especially phosphorus, an essential nutrient for life. As phosphate it is a major component of DNA, cell membranes, and a host of other vital components. Unfortunately, phosphate ions usually exist bound onto soil particles and are inaccessible to the plant. If a gardener scatters phosphate fertilizer on his or her lawn, it will probably still be there in the same top few inches fifty years later, its insolubility preventing it from being taken downward through the soil, even after heavy rain. A root in the soil quickly exhausts all the phosphate immediately around it and can obtain more only very slowly.

Mycorrhizal fungi, exploring the soil far more penetratingly than a root can, greatly enhance the ability of the associated root to take up mineral nutrients such as phosphate. Fungi, meanwhile, may be able to draw nutrients produced by the plant, and maybe even tap the sugars via their intimate connection with root cells. So this is no one-way exchange. Mycorrhizal fungi may take between a tenth and a third of a plant's photosynthetic production in this manner.

The powerful potential of the mycorrhizal associations for resource acquisition may explain why over 90 percent of plants have domesticated their own species of fungus and why, as a consequence, mycorrhizae are the most abundant types of fungi on Earth.

A mushroom found next to a coniferous tree will often be the reproductive structure of a mycorrhizal fungus such as the edible mushroom *Boletus.* The Disney-style red-topped toadstool that is found next to birch trees, called the fly agaric, also betrays an intricate underground association. These two species of fungi belong to a large group called ectomycorrhizae, which make the fine roots of the trees thicker and more stubby. The toadstool is a short-lived reproductive structure, yet the network that produced it often persists long after the tree has died. Herbaceous plants and most trees have different sorts of fungal structures. These are endomycorrhizae, which are so small that their presence or absence can only be detected under the microscope. They get their name because their connections to the plant occur on the inside of the root, whereas in the ectomycorrhizae this exchange occurs nearer the root's surface.

Fungal hyphae can forage through the soil, finding and extracting nutrients while incurring only one-hundredth of the energetic cost of a plant engaged in comparable root building. A single interconnected network of fungi is often shared between several different plants, thus forming a living bridge through which resources may be exchanged from one to the other. The resulting fungus-mediated networking of different plants is as important for the plants as it is for the fungus.

Mycorrhizae are not the only ways in which fungi have engaged in intimate liaisons with plants during their evolutionary history. Many plant cells, not just roots, turn out to contain fungal tissue. Once researchers interpreted such fungal presence as a sign of pathological infection. But the more plant microbiologists have examined the strange guests, the more they have concluded that these internal associates seem to aid plant growth. Internalized fungi have been studied in around a fifth of grass species, and they have been found to confer properties such as drought tolerance and resistance to pests such as aphids, grasshoppers, weevils, and leafhoppers.

One of the more dramatic examples of the advantages for a plant of nurturing a fungal infection throughout its tissues is provided by tall fescue grass, which dominates many of the world's prairies, including those of the eastern and central United States, where it covers around 35 million hectares.

Since the 1930s U.S. agronomists favored a particularly

vigorous variety of tall fescue. For years they had wondered why this grass was so much more vigorous than other varieties, even in drought conditions, and why the grass was unpalatable to cattle. Finally, in the 1960s, its secret was traced to a symbiotic fungus, which was named *Acremonium,* after the acrimonious effect it had on farmers whose cattle were forced to feed on the indigestible grass.

The *Acremonium* fungus grows in the spaces between the cells of the green leaves of fescue. It also confers resistance to drought and increases the number of seeds the grass can produce compared with its neighbors. Seeds are automatically infected with the fungus via a tiny hyphal thread that wraps around the grass embryo while it is still in the seed coat. These seeds sprout into larger and healthier seedlings than do their uninfected counterparts. Not only cattle, but plant-eating insects and root nematodes, too, are put off by the toxic alkaloid compounds present in the fungal cells.

Far from being an occasional quirk, symbiotic fungal infestation is probably the normal state for tall fescue and other grasses. When farmers have tried to grow fields of fungus-free fescue for their cattle, they have found the grasses quickly invaded by the more vigorous fungal variety. Yet this extraordinarily significant and common symbiosis is largely invisible. Only one phenomenon gives an obvious sign of the enhancing effect plants and fungi have on each other's growth: the curious circles known as fairy rings.

Until about a hundred years ago, these circles stirred up excitement among scientists and nonscientists alike: rings of tall, dark grass that seem suddenly to appear as if by magic in parks, playing fields, and meadows. A common sight from the air, fairy rings can be up to nine hundred feet in diameter and four hundred years in age. Each ring is not static, but increases in diameter by up to a foot and a half per year. Sometimes toadstools appear just inside the ring, or the rings overlap, to resemble links on a chain. Since the Middle Ages farmers believed them to mark the paths of dancing fairies. Holding hands in a circle, the spirits danced only on moonless nights, hiding themselves from human eyes.

We now know that a force far more down-to-earth than spirits is at play. Fungi—not fairies—are responsible, not just for the white or brown toadstools, which surface sporadically, but for the dark grass rings themselves. Resources leaked by the fungi to the root systems of each tuft of grass provide a burst of nutrients, such as nitrates, from the fungi's underground activities. As the fungus radiates outward, colonizing and decomposing ever more organic matter in the soil, so the wave of released nutrients turns an ever wider circle of grass tufts a deep, rich green.

Beneath every garden lawn lies such a fungal energy network, one more powerful than a nationwide electricity grid. The human world may be proudly wiring itself up to the Internet, yet fungi were the original networkers of nature.

Think of a whale in the ocean. We see its huge bulk only when it comes up for air. Most of the time, it swims hidden from view, scouring the seas for food. Now think of a toadstool that appears under a tree in your local wood. With it, a fungus is also coming up for air—not to breathe like a whale, but to take its offspring to new sources of food. Masquerading as an independent entity, a mushroom is thus merely a spore-dispersing device for its real body underneath the soil's surface, a huge cotton-wool mass of branching tubes called hyphae that can spread for hundreds of feet in every direction. They seize water and nutrients from the environment and form both the fungus's workhouse and its diffuse equivalent of a brain. Like the worst Internet addict, almost every fungus is permanently online with another member of its community. Loneliness is not a word in the fungal vocabulary.

Myron Smith, an ecologist at Toronto University, has identified one mammoth fungus that spreads its tentacles under fifteen hectares of virgin forest in Montana, weighs one hundred metric tons, and is more than one thousand years old. Such monstrosities may be more the rule than the exception, Smith suggests. At around half a metric ton each, fairy rings are rather weightier than their name and appearance suggest.

Plant fungus interdependence on this scale has profound biological implications. For a start, the networking and resource-sharing abilities of mycorrhizae seem to have created a social security system for plants.

Over forty years ago, Swedish botanist Erik Björkman provided the first evidence for the connective capacities of mycorrhizae when he injected radioactive glucose into the trunk of a Norway spruce tree and then measured it being transferred to a neighboring plant. Despite his suggestion that mycorrhizal fungi were responsible for this dramatic effect, few understood its potential significance. Björkman's findings seemed to fly in the face of all conventional evolutionary theory, in which individuals compete to get the maximum for themselves from a common pool of resources, thereby depriving others—the principle of "the survival of the fittest."

Björkman's colleagues were initially reluctant to grapple with the complex natural communities outside their laboratories. Instead they tested his theory by using whole or partially dismembered plants, grown in laboratory test tubes. This approach rarely allowed them to draw conclusions that held up in the natural world outside the lab. The key to the significance of Björkman's experiments was the extent to which his results suggested trees might be able to overcome variations in resource availability over space, time, and species.

In the early 1990s a team led by a mycologist at Oregon State University, Suzanne Simard, began a series of outdoor experiments in real forest ecosystems in order to measure the resource transfers between trees via their mycorrhizae. Her team found that one fungal mycorrhiza was able to connect together not only many trees of the same species, but even trees of different species. Looking at the network that connected birch and fir trees, they found that they shared up to ten fungal symbionts. To their surprise, the sunlit birch trees seemed to be subsidizing the more shaded fir trees with sugars via their joint network of mycorrhizae.

In their exploration of this wood wide web, Simard's teams have found what could be a new principle of dynamic underground interdependence. Shaded plants, many of which are young seedlings struggling for light, are subsidized by those already bathed in sunshine at the top of the forest canopy. There seems to be an equalization process going on underground. Supplies are shared both within and among species: to those without shall be given, and those with plenty shall have it taken away.

Studies of mixed-species forests on the West Coast of the United States have led David Perry, a soil ecologist also at Oregon State University, to suggest that the sharing of fungal symbionts between groups of trees is so important that they constitute a superorganism, or "guild." This higher organizational level could be one in which important new ecological and evolutionary processes can occur. Any single guild member is able to maintain the fungal symbionts required by all.

Environmental conditions in the forest ecosystem can change rapidly. Yet landslides, hurricanes, or fires may produce effects that a particular tree species finds conducive to rapid recovery. Thick-barked conifers survive ground-level fires better than deciduous trees. So if such a fire occurs, conifers will naturally dominate the next generation of trees. If the fire merely spreads by running among the interlocking branches of trees, however, deciduous hardwood species sprout up from underground roots or seeds more quickly. The foraging strategy and resulting growth patterns of fungi can change completely depending on the resources available. Similarly, multispecies fungal guilds stand to gain overall if they can supply a variety of tree species, each of which can take the best advantage of an opportunity created by a different kind of environmental disturbance. By buying all the raffle tickets in the pack, each fungus guild ensures that whichever number is drawn out, one of its members will be the winner.

If you have ever tried to mow a lawn in neat stripes you will have probably seen this dynamic equilibration process at work. Perhaps because of an uneven surface, a badly adjusted mower, or an inattentive human operator, garden lawns often end up with some strips of grass taller than others. A dip may leave some tufts higher than those around them, or a grassy hump could have been cut too short. However, when you come back to mow it again two weeks later, a subtle transformation has taken place. Despite different patches' being left at a variety of contrasting heights, the grass all over the lawn is now, give or take a few fractions of an inch, the same height. Perry's experiments suggest that this is the product of a mycorrhizal welfare state and that it could be the rule for every lawn and woodland.

Only one other microbe can rival the power of fungi to form a dynamic duo with plants. This group of bacteria, called rhizobia, is probably the single most productive family of organisms on the planet. Many plants manage to obtain one range of nutrients from their mycorrhizal fungi, but it is the legumes that can make the most of rhizobia's unique innovation—the nitrogen-providing nodule.

The second largest plant group, legumes include not only hundreds of wild species from acacia bushes to vetches, but also familiar crops such as peanuts, garden peas, alfalfa, and kidney, soy, haricot, and almost every other sort of bean. That the four-leaf clover is a sign of good luck seems fitting because the evolutionary good fortune of clover and its fellow legumes has long been suspected. Writing in the fourth century B.C., Theophrastus provided the earliest description of how legumes seemed to "reinvigorate the ground." Aristotle also commented on their effects, but neither sage knew why.

Grow a nonleguminous crop next to a legume and it will usually grow better than if it were grown close to a member of any other plant family. Crop rotation practices usually include the growth of a stand of legumes that is plowed back into the soil after harvesting. Theophrastus himself summarized the effect: legumes "are not a burdensome crop to the ground," he wrote, "they even seem to manure it." Many gardeners describe legumes as "living manure" because of what we now know to be their ability to take nitrogen from the atmosphere—a process known as nitrogen fixation.

Present in the proteins and DNA of all organisms, nitrogen is essential to every living process. Yet despite comprising 80 percent of the air, its gaseous form is inert and unavailable to plants, fungi, and animals. Animals usually obtain nitrogen by eating other organisms, as do carnivorous plants. Most other plants rely on microbes that can transform nitrogen into a digestible form, or—as in the case of legumes—they form a symbiosis that enables them to obtain it directly from the air.

It was not until the end of the nineteenth century that biologists at last understood the microbial mechanism behind the beneficial effect that legumes had on the growth of other plants. The secret was in the pink, lumpy nodules that are attached to their roots. Victorian microscopists observed that the nodules were full of dark-staining dots, unlike the usual contents of plant cells. In 1885 Martinus Beijerinck—a Dutch microbiologist, student of de Bary, and pioneer of the study of the contagions that we now call viruses—described how a group of bacteria organized itself as if it were a multicellular organism and collectively invaded the roots of leguminous plants.

Beijerinck proposed that in response to these bacteria, legume roots form nodules, which redden because of the presence of a protein that in detail resembles the hemoglobin in our blood. Having infected the root, the bacteria live inside the nodule cells, taking nitrogen from the air and "fixing" it, thereby providing nitrates to the plant and surrounding soil. This group of bacteria was soon named rhizobia, simply meaning "root-living."

Like many of the bacteria that colonized the early oxygen-free Earth, rhizobia are poisoned by too much oxygen. Via a complex joint metabolic pathway, the rhizobia and plant together synthesize a legume version of hemoglobin that turns red, just like our own variety, as it binds to oxygen. But instead of transporting oxygen, legume hemoglobin acts as an oxygen blanket—trapping the oxygen and so insulating a plant's rhizobial associates from being poisoned by it.

Nitrogen fixation is rare in nature because it involves the breakage of the triple linkage between nitrogen atoms, one of nature's strongest chemical bonds. To perform this energy-consuming task, the legume supplies copious amounts of sugar. This, together with the oxygen protection from their hemoglobin, enables rhizobia to fix nitrogen for the plant-bacterial complex remarkably efficiently.

A process of gene transfer has stabilized the association between the rhizobia and its legume host. Each time a new legume seed germinates, it starts a molecular conversation with the rhizobial symbionts that have lain dormant in the soil. Each associate recognizes the other, and the rhizobia burrow through the plant cell wall. Together, the two symbionts construct the specialized nodule tissue and the necessary hemoglobin. The rhizobia, trapped inside the nodule, can no longer divide and reproduce. In this cyclic association, each new generation of plants is thus reinfected by rhizobia from the surrounding soil.

By fixing nitrogen, rhizobia create a positive feedback, leading to an increase of their own populations and, under most conditions, the growth rate of the legume. The more nitrogen fixed for the plant, the more energy, in the form of plant sugars, becomes available for new growth and the more nodule production, leading to greater numbers of rhizobia, fixing ever larger amounts of nitrogen.

The rhizosphere is the name for a root's sphere of underground influence. Plant roots do not only leak nutrients, but also actively secrete sugars and amino acids. By providing food and energy for free-living bacteria, protists, and fungi, they receive, in return, mineral nutrients.

The secretion of a nutrient-rich jelly called mucilage by the growing roots makes for a compost heap of organic matter that stimulates the growth of bacteria that break it, and soil detritus, down. Catalyzed by the plant, this decomposition thus releases further nutrients in a form the plant can use. Protists, such as amoebae and ciliates, roam the rhizosphere, grazing on the huge numbers of rhizosphere-feeding bacteria. When the protists digest their prey, they release more nutrients, such as ammonia. This nutrient is also released by both mycorrhizal and nonmycorrhizal rhizosphere fungi, which join the bacteria in the root's compost heap.

Other rhizosphere inhabitants often keep in check certain species potentially harmful to plants, such as pathogenic fungi and nematodes. So long as they are prevented from infecting the root, soil fungi and nematodes can actually increase the growth rate of the plants by adding to the general turnover of nutrients in the rhizosphere, like flies and their maggots enhancing the recycling activities of a compost heap in our own gardens.

In temperate climates, the rhizosphere is at its most productive in the late spring and summer, when the soils are moist and the temperature is high. At these times the roots advance into new areas of soil, turning barren earth into a productive resource. The tip of every root contains tissues that divide continuously, each at a rate of up to twenty thousand new cells per day. The fluid produced by these cells is often exuded in such high volumes that small droplets form at the end of each root.

The slime from the root tip is left behind as the root grows through the soil, acting as a lubricant against sharp soil particles, but also preventing the rhizosphere from drying up. Without water the bacterial teams would have no medium in which to eat and respire. The mucilage ensures that a continuous film of nutrients and water covers the soil particles and the root, thus keeping the root's minute compost heap in action. In contrast to mycorrhizae, which are plumbed directly into plant roots, here at the root tip, the plant's cells cream off nutrients from the diverse activity of an external microbial community, but are only indirectly connected to them through the jellylike mucilage.

Given enough sunlight above ground and water below ground, this positive feedback can continue so long as there is new soil to explore. As the root grows, the burgeoning rhizosphere ecosystem expands. The more nutrients acquired from the rhizosphere, the more root cells have food and energy to continue their rapid growth.

Plants, the first gardeners of Eden, act as a catalyst to the formation of widespread associations between organisms that perform synergistic metabolic tricks. Mycorrhizae epitomize the mechanism by which the soil-atmosphere interface has been conquered by alliance builders. Plants are masters of the complementary skills of gas exchange and photon collection, while their symbiotic fungal foragers are uniquely suited to exploring a three-dimensional environment and have a huge potential to weather rocks, release nutrients, and break down dead organic matter, thus compensating for the few nutrients available in most soils. If a freak virus were able to eliminate plant-associated fungi tomorrow, every grass, bush, and tree would soon shrivel and die.

Biologists who maintain that fungi that live inside

plants are pathogens do not do so entirely without basis. When plants eventually do die—be they fallen leaves from trees or annual flowering plants at the end of the summer—internalized fungi get the first bite at the corpse. More significantly, if a plant is injured or malnourished, these same fungi can exploit its weaknesses and transform themselves into pathogens, driving the host to an early grave. However, such pathogenic relations are the exception. Healthy plants can usually tame the same species of fungal and bacterial symbionts that can be a liability in less healthy specimens.

Even the fungus-like microbe implicated in the Irish potato famine is frequently present as a symbiont in hundreds of potato species across the world and causes no blight epidemic. The disastrous outbreak in Irish potato crops was the product of a variety of fungus from South America suddenly coming into contact with a new variety of potato that it had never encountered, together with the nutrient-deprived state of the potatoes in Irish peasants' fields.

David Thurston, a plant pathologist at Cornell University, has studied the traditional practices of potato cultivation in the South American Andes. Though poor by our standards, indigenous Andean farmers have grown potatoes for thousands of years apparently without a single famine. Thurston's research suggests that this success stems from agricultural systems that reuse waste products to maintain their rich soil and that exploit a generous mix of potato varieties. Instead of being forced to wrench ever higher yields from a single variety of potato grown on a fixed amount of land, as the Irish were, Andean farmers have devised a system that keeps their potatoes nourished, and reliably productive. The crop is healthy enough so that any fungi it encounters can be either fended off or tamed.

De Bary's legacy to biology was his demonstration that microbes can exist at many points on a symbiotic continuum from nutrient provider to agent of disease. The potato famine's legacy was its demonstration that sociological, economic, and ecological conditions can combine to shift a bug from one end of the continuum to another. For by allowing the Irish peasants to be pushed onto the poorest, often water-soaked land and thereby leaving their crops vulnerable to disease, the British administration of the day, not the fungi, was the real cause of the blight.

CHAPTER 3

HIDDEN GARDENS of ATLANTIS

The great depths of the ocean are entirely unknown to us. Soundings cannot reach them. What passes in those remote depths—what beings live, or can live, twelve or fifteen miles beneath the surface of the waters—what is the organisation of these animals, we can scarcely conjecture.

JULES VERNE, *TWENTY THOUSAND LEAGUES UNDER THE SEA*

Simon Schwendener announced his bold hypothesis on the dual nature of lichens in 1869. The following year the French inventor of science fiction, Jules Verne, completed his fantastic marine epic, *Twenty Thousand Leagues under the Sea,* in which his intrepid explorers, led by Captain Nemo, chart the depths of the oceans by submarine. While Schwendener's theory was sucked into a highly charged controversy that was to claim the scientific careers of Beatrix Potter and a host of her fellow pioneers over the coming decades, Verne's tale of underwater adventure and discovery was becoming his most successful. Today it is regarded as his most prophetic work.

Verne crafted his adventure story after careful research. The design of the submarine that carried his heroes on their journey under the oceans was inspired by drawings of local inventor Jacques-François Conseil, and reports of prototypes such as the *H. L. Hunley,* used by the Confederates against the Union navy during the American Civil War a few years earlier. Verne also picked up enough oceanography to provide biological realism to his tale, even though he was an engineer by training and the scientific study of the seas was only just beginning.

Among the highlights of Captain Nemo's voyage were his encounter with a giant squidlike monster and a glimpse of the lost city of Atlantis, a "Pompeii buried under the waters." Writing a century after the book's publication, Verne biographer Peter Costello remarked that the science of the deep sea had hardly progressed in the intervening years. In at least one way Costello was mistaken. Within a decade of Nemo's voyage, squid and fish in the ocean depths were discovered to glow. It meant that Atlantis would have had its own biological street lighting.

In 1890 one of Verne's fellow countrymen, biologist Raphael Dubois, proposed that the source of the squid's luminescence was its symbiotic bacteria. He also extended the explanation to some beetles that glowed in the dark. For the following thirty years his remarkable hypothesis was the subject of excited, but largely sympathetic, conversations among his colleagues. They inspired *Les Symbiotes,* a book by Paul Portier, which provided the first comprehensive exploration of symbionts in the natural world. However, when Portier's book was published in 1917, it became the object of ridicule by the most influential biologists in France, the United States, and Britain. Fearing disastrous consequences for his career if he remained silent, Dubois retracted his previous theory, giving a completely different explanation of his original observations. We now know that his original intuition was correct; bacteria are at the heart of the glowing colors of many animals on sea and land.

In 1977 scientists from the Woods Hole Oceanographic Institution, in Massachusetts, made a Verne-like voyage in their newly built space-age deep-sea exploration vessel, *Alvin.* Like Captain Nemo's crew, they came upon a menagerie of animals from the ocean abyss that nobody had ever seen before—crabs, willowy anemones, bright-red shrimp, and symbiotically luminous fish. Ninety-five percent of the creatures seen on the trip were later declared to be new to science.

Imagine three of us are crammed together in *Alvin*'s metal capsule, about the size of a broom closet, as we descend through thousands of feet of water. The first thing to strike us about the deep sea would be how utterly black it is. Whatever time of day, season of the year, or weather condition at the surface, at a depth of one and a half miles it is always pitch dark. Outside, everything is still and silent. As your eyes become used to the eerie environment, they start to perceive some of its secrets.

Every hour or so, one or two faintly glowing particles drift down the water column from above. As these specks of light glide slowly through the blackness, a small fish with grotesquely enlarged eyes appears. It seems to have spotted these tempting morsels and tries to swallow one of them.

At the moment when this goggle-eyed scavenger reaches what had seemed to be an appetizing snack, a huge, foot-high pair of jaws appears out of nowhere and snaps shut like a bear trap. The big eyes and their owner disappear, leaving only the huge jawed predator. This scene of carnivorousness might seem a far cry from the world of microbial liaisons, yet symbiotic bacteria scripted it by emitting every ray of light. Only in the past few years have researchers begun to uncover the biology behind remarkable dramas such as this one that in the abyss are an everyday occurrence.

Ultimately powered by sunlight, most biological activity in the oceans happens at the surface. Very few nutrients reach the extreme depth of the ocean floor. Occasionally, however, particles of organic matter, like the ones we saw passing *Alvin,* do reach the bottom. What began as a pellet of waste products from an organism such as a shrimp floating near the surface will be scoured for nutrients on its passage down. It will also be colonized by a variety of bacteria and other microbes called protists, which chemically detect such particles from a distance and gather on them, forming a miniature sinking consortium. When oxygen is at low enough levels, some of the decomposing bacteria emit light.

These bacteria thereby confer a slight glow to many of the detritus particles that reach the depths. The descent of these morsels is so infrequent, and they are so small, that the ocean bed remains severely lacking in sources of nutrition. Deep-sea fish are consequently widely dispersed and have huge eyes so as not to miss a single particle.

Enter the predators. The divas of the deepest sea are undoubtedly the carnivorous females of anglerfish species that can be up to two feet long and a foot high, this height being almost completely taken up by a huge jaw. When food becomes available they must be able to eat it, however big it is.

Large mouths are common down here, but the way these anglers catch their fish is not: they do it by light fishing. A rod extends from their equivalent of a forehead. On its end is a glowing bulb-shaped pod of luminescent bacteria. The anglerfish are thought to attract their prey because their bulb mimics the appearance of a rotting detritus particle. Lured to take a bite, the innocent grazer is swallowed whole by the anglerfish. Predators skilled at bacterial cultivation have thereby created a means by which they can attract potential prey.

The process that led to the eventual incorporation of

luminous bacteria into many marine animals seems to have started with lunch. For many millennia fish and squid regularly ingested a consortium of different glowing bugs. Rather than being eliminated by their digestive systems, many bugs can survive in these guts. Some of these bacteria would be able to increase their populations by feeding on the nutrients concentrated there. It may even be that luminosity first evolved among bacteria to exploit the positive feedback that could result from passing through the animal digestive tract. The brightest bugs would be eaten most often and thus be able to increase their populations by the greatest amount.

Taking our evolutionary scenario a step further, imagine that one population of luminous bacteria evolved the ability to persist in a long-term liaison in a fish gut, rather than be defecated. The glowing cargo might just enable the fish's large eyes to make out desperately needed food morsels, and over time there would be strong evolutionary pressure for the association to become more effective. Fish whose luminous bacteria had colonized the region near their eye might catch more prey.

But since luminous bacteria leave no fossil record, attempts to reconstruct their evolutionary history, however plausible, must remain speculative. Whatever the case, luminosity has evolved in virtually every large animal group in the ocean—from the jellyfish to annelid worms to crustaceans, even to the eyeless starfish.

Starfish, like many smaller marine organisms, have little chance of rapid escape from a predator. Some seem to have evolved a strategy of emitting a flash when threatened, thereby perhaps startling their attackers. This same behavior has been extensively studied in protists such as the charmingly and aptly named *Noctiluca.* It lives near the surface of the open ocean and is grazed upon at dusk by small shrimps. But by using a bright flash, *Noctiluca* seems to dazzle an approaching shrimp, gaining it enough time to escape.

If *Noctiluca* can flash brightly enough to illuminate its attacker, it may also make the shrimp visible to a hungry passing fish—the so-called burglar alarm effect. A shrimp needs not only to eat, but also to avoid being eaten. The daily migration of copepods to feed at the surface just before it gets dark may be a response to the effectiveness of this flashing alarm strategy. The shrimps must feed when there is too little daylight for fish to see them, but before the luminous alarm from *Noctiluca* would leave them exposed to attack.

The foot-long flashlight fish, *Photoblepharon,* has yet another symbiotic innovation. It has evolved biological headlights—pouches of luminous bacteria next to its eyes that send out two bright beams into the blackness and help it catch prey. They may be used as a signal that allows individuals to group together in schools. And unlike most headlight-bearing species, *Photoblepharon* also has a ingenious flap to cover the headlights when they are not required, perhaps when there is a predator around. It is truly a Ferrari-fish. Some species of fish have even evolved the pouches for communication, sending out long wavelengths of light that cannot be detected by other species.

Symbiotic relationships also exist in shallow waters and in creatures far more alien than those spied by *Alvin.* The Cambrian explosion 550 million years ago was the period that saw the abrupt appearance of the skeleton-bearing ancestors of today's animals, plus many more types that later became extinct. Yet around 50 million years before this biodiversity boom, the Ediacarans appeared and blossomed into a host of even more extraordinarily shaped, soft-bodied, multicolored creatures.

Named after the Ediacara Hills in South Australia, where they were first found, Ediacaran forms are unlike anything else in the fossil record. One, *Pteridinium,* resembles the tread of a bicycle tire, while *Dickinsonia* looks a bit like chewing gum with the imprint of a boot heel upon it, and *Cyclomedusa*'s shape has an uncanny similarity to a fried egg. Some were unquestionably preskeletal animals. Others have been so difficult to interpret that even as late as the 1960s, biology textbooks were being published with pictures of them with legends such as "Mystery fossil: one of the oldest ever found; no known relationship with any other creature—living or dead."

The Ediacarans soon colonized the seafloor over much of the globe, towering over the early wormlike animals that shared the same sandy-bottomed marine habitat. Their body forms were similar to those of modern jelly-fish and mollusks, and they were even equipped with nervous and sensory systems. Yet they were not mobile and stayed attached to the shallow seafloor like rows of Post-it notes.

Initially it was assumed that Ediacarans could be shoe-horned in with other animals, just as lichens had been lumped with other plants a hundred years before. Then, in 1982, Adolph Seilacher, a paleontologist at the University of Tübingen, in Germany, reinterpreted the group as belonging neither to the animal nor to any other kingdom, but to quite another realm.

Three shared features of the Ediacarans may mark them out as a unique group. First, their whole bodies were inflated by water. Second, all members of the group were characterized by vanes or lobes composed of repeating segments. Rather than developing from a single blastula cell, as animals do, the basic unit of Ediacarans is a cell family—a kind of intermediate stage between mere clumps of individual cells and fully organized tissues. Third and most strikingly, they fed—to all appearances—not on each other, but on the products of their internal microbial associates.

The Ediacarans may have been unique in the history of large, complex life in having no predators. There is nothing to suggest that they possessed teeth, claws, snouts, or any appendage that might have allowed them either to feed on their neighbors or defend themselves. That is why Mark McMenamin of Mount Holyoke College in Massachusetts, for one, loosely compares the Ediacaran world with a biblical Eden; his Garden of Ediacara hypothesis proposes that this was a war-free realm in which peaceful residents usually made their own dinner, via photosynthesis, rather than by attacking and lunching on others. To the extent that they derived their nutrition by containing symbiotic photosynthesizers, Ediacarans can be considered to have been plant-animal hybrids.

This ancient realm of life not only evolved strange and unique growth forms. They also, McMenamin suggests, possessed metabolically transforming microbes as a keystone to their existence. He has found fossil evidence that he interprets as suggesting that Ediacarans had sensory structures sophisticated enough to be regarded as crude brains. They behaved as plants that were half animal, or else animals that were half plant. They preceded both, but were probably obliterated before any organism ever evolved into either.

Though none of the Ediacarans survived until the present day—they vanished after only 50 million years—an extraordinary creature has recently been described that seems to provide a modern analogue. A green jelly cylinder of up to a foot in length has been found living in the peaty bogs of Massachusetts. Though it lounges at the bottom of the bog pools across the world, it has been very hard to study.

First described from bogs in East Anglia, England, in 1786, the jelly was named *Ophrydium,* after the Greek for eyebrow, which its tiny constituent cells resembled. It remained virtually ignored by biologists for two hundred years, being of no known economic or medical importance. Yet when a team led by Brian Duval of the University of Massachusetts at Amherst analyzed some local samples from nearby bog pools, they were struck by the similarity between its organization and that of the Ediacarans.

The *Ophrydium* cell, a cilia-covered protist, is less than a millimeter long, yet it is surrounded by a translucent jelly. Each jelly cylinder, normally the size of a hamburger bun, is made up of hundreds of cells arranged in ordered rows in a three-dimensional matrix. Each cell contains symbiotic protists that give the whole its green color. What is most surprising about this blobby photosynthetic factory is that it is also the permanent home for a score of other completely unrelated organisms, including the protist well known to every high school student, *Paramecium,* several small freshwater crustaceans, flying-saucer-shaped algae called diatoms, and even worms.

The rich community living inside the *Ophrydium* jelly mass is no random assemblage; together they form a coherent, integrated mass. Some biologists, such as Lynn Margulis, and paleontologists, like McMenamin, believe *Ophrydium* to be the best living analogue of the Ediacarans yet found. They speculate that *Ophrydium*-like organisms lived before what we now know as animals finally took off. But their theory is as yet unproven, since no *Ophrydium* fossils have been found.

Gregory Retallack at the University of Oregon has a very different theory about the nature of Ediacarans. He believes they were underwater lichens. Rather than being an experiment in the early evolution of life, Retallack believes this marine group went on to become the lichens we see around us today. Contrary to McMenamin, Retallack believes these creatures were producers, not consumers, of oxygen, and that the Garden of Ediacara was in fact the Age of the Lichens. Unlike the age of the dinosaurs, which ended abruptly with their elimination, the lichens vanished from one environment, the sea, only to become spectacularly resilient colonizers of the land.

A scientist with a computer-generated simulation of a past world is worth a thousand tables of dry data, and McMenamin and Retallack both use their digital image projectors to dramatic visual effect when illustrating their vision of what life was really like on the floor of the ocean 600 million years ago. But no matter how persuasive the arguments of each are, without fossil evidence we will never be sure which of their two worlds, if either, really did exist.

One group of shallow-water animals for which there are extensive fossil records are the clams and mussels—a group collectively known as the bivalves. Generations of scientists had assumed that bivalves, which evolved during the late Cambrian period, were almost exclusively filter feeders, their thin, delicate, translucent gills, much beloved of seafood eaters, directing tiny morsels of food toward their digestive systems. However, a recent reexamination of both the live and fossil bivalves by a team led by Daniel Distel, a microbiologist at the University of Maine, has revealed that many of them do not filter feed at all. Instead they obtain their food from internal microbial associates.

Unlike the photosynthesizing protists present in *Ophrydium* cells, which use light energy to make food, these shellfish contain ancient chemosymbiotic bacteria. The bugs metabolize hydrogen sulfide—the rotten egg gas—and carbon dioxide, chemicals that are normally unavailable as energy sources to animals. This is as different a metabolism from most other animals as it is from plants, yet it seems to have been a common feature not only in widely distributed bivalve families, but also in other mollusks, annelid worms, and nematodes. The strategy seems so successful at exploiting certain specialized ecological niches that their anatomy and physiology has remained unchanged for around half a billion years. The contribution of the alliance to the metabolism of the chemosymbiotic bivalves is attested to by their abnormally large size. While the mussels we may buy at the seaside or eat in a restaurant might be a couple of inches from end to end, the symbiotic bivalve *Bathymodius* can reach a foot in length.

Congregating in characteristic flower-shaped lobes that have led one study site to be nicknamed "the rose garden," chemosymbiotic bivalves develop special organs to house their bacterial symbionts, while the normal digestive system that in nonsymbiotic bivalves extracts nutrients from the filtered food is either much reduced in size or completely absent. Following the discoveries by Distel and his colleagues, bivalve biologists and paleontologists have been forced to go back to previous specimens, often to find that their mode of nutrition had been misinterpreted. Gut-less mussels that had previously been assumed to have digested their food outside their bodies can now clearly be shown to follow this alternative mode of nutrition. Rather than eating other organisms, like most animals, they allow their symbionts to make their food out of raw chemicals, more like green algae or plants.

Chemosymbiotic bivalves have now been found in a wide variety of low-oxygen environments, where a high level of hydrogen sulfide excludes most other organisms. These include sea-grass beds, sewage outfalls, rotting whale carcasses, and even the decaying cargo of a sunken coffee freighter. But they were first found near deep-sea hydrothermal vents.

These vents are areas of the seafloor where lava bubbles up from the earth's molten core to contribute new rock to the planet's outer crust. Around two miles beneath the ocean surface, they are the locations of ocean-floor spreading—the driving force behind the gradual drifting of continents over geological time. The jets of black, sulfurous, magma-heated water emitted from the vents send up conelike "black smoker" chimneys as they precipitate the minerals they bring with them. Some researchers believe that life itself began down here, with temperatures of up to three hundred degrees Celsius, little oxygen, and no light. Today, microbial communities in such regions grow at a density of one hundred thousand cells per gram of rock, and although little is known about the life that exists in the dense, solid rock of the chimney walls—only a few hundred of the many thousands of kinds of microbes found in association with the chimneys have been isolated, and only a handful of the three hundred species are visible to the naked eye—the vents nonetheless offer us one of the most graphic examples of how alliances with microbes have allowed animals to exist in the most extreme environments on Earth.

On July 5, 1998, a team led by Edmond Mathez of the American Museum of Natural History in New York raised a section of a black smoker for the first time ever. The chimney, when pulled to the surface by a ship lying two hundred miles west of Seattle, looked like the trunk of a giant redwood tree, festooned with pink tubes that must be one of the planet's strangest looking marine organisms, a chemosymbiotic tube worm that belongs to a group called *Riftia*.

Like rows of living fire brigade hoses, whose bodies are

about ten inches wide and six feet long, *Riftia* sways in the convective currents that bring it sulfurous chemical energy from the bubbling vents. Unlike its nonsymbiotic relatives elsewhere in the ocean, *Riftia* lacks both a mouth and a gut, and it had been unclear on what it fed until Harvard University microbiologist Colleen Cavanaugh and her colleagues came upon telltale sulfur crystals inside the tubes. Each worm, Cavanaugh found, contained populations of domesticated sulfide bacteria, very similar to those inside bivalves.

Riftia's metabolism has had a huge impact on thinking about the hydrothermal vent ecosystem. Biologists used their new understanding of this remote habitat to take a second look at ecological processes nearer the seashore. Taking into account the bivalves, as well as free-living bacteria, they found that up to 90 percent of all decomposition of organic matter in coastal marine sediments may be due to the same sulfate metabolism that occurs in *Riftia*'s chemosymbionts. So having thought *Riftia* was a freak, biologists are now discovering that bacterial symbioses allow a range of extraordinary marine metabolisms.

Shipworms, the bane of every wooden ship from ancient Egypt to the Battle of Trafalgar, are tiny bivalves, not worms at all. Unlike most bivalved shellfish, in which two calcified plates enclose the whole animal, the shipworm's plates have become toothed rasps that they use to bore into wood. Wood may not look very tasty but it is a rich source of bacterially metabolized energy. The worm's tiny gills contain bacterial symbionts, which break down the wood's lignin as if they were digesting grass in a cow's stomach. But unlike grass, wood is junk food, being very low in nutrients, especially nitrogen compounds. So most remarkable of all, the shipworm hosts additional kinds of bacteria to provide these nutrients as well.

Having found the marine world awash with symbiotic interactions, a number of biologists are turning their attention to how such associations interact with animal patterns of reproduction and development. Working at the University of Hawaii, marine biologists Margaret McFall-Ngai and Ned Ruby have carried out just such a study of the symbiotic light-emitting organ of the bobtail squid.

Though a convenient research subject, being both abundant in shallow coastal waters and easily kept in seawater aquariums, the bobtail squid is a no less dramatic beast for that. Its use of light almost rivals the spectacular image seen from *Alvin* below. Many squid and fish species live near the ocean surface. To a predator underneath them, their bodies form fortuitous silhouettes, casting identifying shadows against the downwelling light. Even at night, when most of these animals and their predators rise to feed, any moon- or starlight has the same effect as the sun. Many fish and squid use light from symbiotic luminous bacteria, projected downward at the same intensity and color as the light from above, and thereby prevent their silhouette being seen.

Though the bobtail is only around six inches long, its symbiotic light makes the difference between getting home for dinner and being somebody else's meal. Yet generations of marine biologists had little idea of the struggles of light and darkness going on underneath their research ships because, like Jules Verne's explorers, they could not conceive of these kinds of phenomena.

Shortly after hatching, in just a couple of hours, young squid start to encourage suitable bacteria by waving water past the pores in their skin where the bacteria must take up residence. Despite the surrounding seawater containing millions of diverse kinds of bacteria, types other than the luminous *Vibrio* bacteria are excluded. The molecular exchange that leads to this mutual recognition appears to be very similar to the one practiced by legumes to exclude nonsymbiotic rhizobia. It's what McFall-Ngai describes as "a really delicate conversation—our bacterium says, 'are you the right squid,' to which the squid replies 'are you the right bacterium,' and they do this several times."

Reproducing with a generation time of less than thirty minutes, the symbiotic *Vibrio* are drawn through ducts, lined with cilia, which are just like the tiny pumping hairs on the inside of our throats. Then the bacteria multiply to fill the dead-end spaces where the squid will make use of them.

Within ten to fifteen hours there are around a million *Vibrio* cells inside one of the pores of the squid. Suddenly, a wave of light passes through the organ and it illuminates like a fluorescent bulb. Further growth in the numbers of *Vibrio* cells is regulated by the squid cutting off the oxygen that normally leaks out of its cells. Any mutant bacteria that are unable to produce light are ejected. In response, the *Vibrio* cells lose their swimming apparatus, which serves no function for the squid. Their sole role now is to reproduce and create light.

As we move out of the shallow water, we find that even within the seemingly lifeless layers of foul-smelling mud at the water's edge, there are rich examples of novel consortia. One of the best, the ribbon-shaped *Kentrophoros*, is a ciliate just a tenth of a millimeter long. Some protists sweep food into their cell mouths with beating cilia, but *Kentrophoros* has none of these waving tentacle-like structures. Nor does it show any sign of ingesting any food from its surroundings. Instead, it carries a dense coating of bacteria on its back. These are normal rod-shaped bacteria, but aligned lengthwise on their tips. In a pattern like skittles at the fairground, these bacteria cover the upper surface of their protist associate.

Two biologists, Bland Finlay from the Fresh Water Biological Association, in the English Lake District, and Tom Fenchel of the University of Copenhagen, found this fascinating association and what they call its "kitchen garden" of bacteria in mud on beaches near Copenhagen.

Having cultivated a crop of bacteria, *Kentrophoros* can simply gather a meal from a small patch of its garden by folding in its outer membrane. This invagination creates a package of bacteria sealed inside the feeder's membrane that can then be digested inside the cell, just as if it were a meal that had come through its mouth. And as our skin heals after a cut, so the membrane seals itself. It then allows the bacteria on its outer surface to divide, replacing those that have been consumed. By a simple fold of its skin, this master harvester can thus enjoy what Finlay calls "an everlasting picnic."

Since it lives in the dark under the mud, the domesticated bacteria living on *Kentrophoros* cannot harness the energy of the sun by being green and photosynthetic. Instead this protist must move its garden to satisfy its rather peculiar dietary requirements. *Kentrophoros* travels to where there is hydrogen sulfide fertilizer for its bacterial crop, together with carbon dioxide and small amounts of oxygen. These three gases occur together in a single thin layer of marine sediment, where the sulfide diffuses up from the oxygenless depths and oxygen diffuses down from the overlying water. In this narrow niche *Kentrophoros* becomes a cornucopia.

More than one hundred thousand protists can live in a cubic millimeter of wet mud, but so few biologists study this habitat that only a tiny proportion of protist biodiversity has been recorded. Among the few that have been closely observed, scientists have found it difficult to locate any protists that do not contain domesticated bacteria. The fertile marine mud in which *Kentrophoros* is found may well be where the complex cells characteristic of all plants and animals originated.

Low-oxygen environments do not just occur in marine mud. Sewage tanks, lake beds, and the deeper layers of humanity's landfill are both oxygen-free and contain protists that are able to carry out a complete life cycle without that element we find vital. They specialize in domesticating at least two sorts of bacteria. Some protists, like those that have been found in a low-oxygen Spanish lake, support thousands of bacteria on their external surfaces just like *Kentrophoros,* except their bacterial coats use sulfates instead of sulfides. Others contain internal populations of bacteria that produce the highly flammable gas methane. In some landfill sites, so much methane is produced that it has to be released and burned off to reduce the risk of underground explosions.

Rather than relying on one crop, some protists can choose which microbe they want to cultivate. One, *Euplotes,* always picks a photosynthetic associate for its internal garden, but not always of the same type. Its usual choice is the ubiquitous but much smaller protist *Chlorella. Euplotes* encloses each cell in a vacuole, as if it were going to eat it, but instead encourages it to photosynthesize by supplying it with ammonium, a vital ingredient of light-harvesting enzymes and other proteins.

Like other microbe domesticators, *Euplotes* must control the proliferation of its garden to stop it from outgrowing its host. In this case *Euplotes* alters the amount of ammonium it provides to *Chlorella*. Instead of synthesizing new proteins for production of their offspring, the *Chlorella* are driven by their lack of ammonia to make sugars. *Euplotes* uses these sugars in its own growth. When *Euplotes* starts to reproduce, it can supply copious amounts of ammonia that will allow the alga to furnish both of the resulting offspring with a well-stocked photosynthetic kitchen garden.

Euplotes does have to make at least one compromise for the sake of its guest. *Chlorella* requires carbon dioxide that is only present in sufficient quantities at depth, but also needs to be near the light to photosynthesize. Just like its ribbon-shaped relative and using the same oxygen-sensing behavior, so *Euplotes* migrates to the borderlands of the light and carbon dioxide zones.

The ocean's most dramatic bacteria farmer is a protist that actually becomes vampiric depending on the weather. Dianne Stoecker, an oceanographer at the University of Maryland, has discovered surface-living water protists, oligotrichs, which feed on smaller photosynthetic protists. But during the long, hot summer days, their behavior changes. Before digesting the prey, the ciliate predators suck out the chloroplast. This fragile organelle is then carefully tended to provide extra food for the oligotrichs until the green slaves eventually wear out and are themselves digested. This trick is performed not just by microbes, but also by organisms as big as a small rodent, such as the sea slug *Elysia*.

While those obsessed with the big and the colorful might dismiss some domestication phenomena as mere microscopic curiosities, they would be hard pushed to belittle the evidence of this large, jade green shoreline mollusk. *Elysia* makes its living by grazing *Codium* seaweed, dubbed "dead-man's fingers" because of its characteristically cylindrical lobes. Instead of biting off pieces of seaweed as a garden slug might devour lettuce, *Elysia* sucks out the seaweed's chloroplasts and harbors them between its own cells for several months. The chloroplasts provide so many nutrients that during the summer, the mollusk can go for three months without eating. And after the first crop dies off and suffers digestion itself, *Elysia* returns for another crop.

For many years it was assumed that *Elysia* simply supplied housing and nutrition for the chloroplasts, but a team of marine biologists from Texas A&M University, led by Cesar Mujer, recently found evidence that the sea slug plays a far more active role. Its cells seem to help their green food factories activate the genes responsible for the production of the proteins that allow their chloroplasts to continue their metabolic activity for such long periods. It's as if *Elysia*'s tiny prisoners are being given the tools to let them work harder for longer. Importantly, in the predator-rich habitat of the seashore, the green guests also act as an unsurpassable camouflage for *Elysia* in its seaweed-dominated habitat.

While almost every example of a marine symbiosis discussed so far has only been discovered during the last decade or so, perhaps the most attractive example has been studied since ancient Greece and was a favorite of Charles Darwin. This is the largest and possibly most spectacular phenomenon of marine life, the reef-building coral—the tropical rain forests of the oceans. These animals may have become the most tenacious members of the animal kingdom because of the self-sustaining circle of positive feedback created by their symbioses.

With the advent of global climate change marine ecologists have become concerned about the "bleaching" of corals, which, they suspect, provides an early warning sign of human-induced global warming. Such bleaching has nothing to do with household-cleaning agents, but rather describes the loss of the coral's symbionts, called *Symbiodinium,* and the consequent fading of its green color. Some coral specialists now believe that this bleaching may be an extreme form of the natural strategy by which corals alter the makeup of their microbial associates in order to cope with a changing environment. Rather than a symptom of decline, it seems to be an evolved mechanism that has allowed corals to outlive less ecologically flexible animal groups. The real issue to be faced is whether the corals can continue to do so, given the speed with which human activity is affecting the reefs.

Robert Buddemeier and Daphne Fautin, coral biologists at the University of Kansas, have found that both corals and *Symbiodinium* can form symbioses with more than one type of associate. On this basis they suggest that bleaching may provide corals with the opportunity to be repopulated with a different type of photosynthetic protist. Some combinations may be less proliferative but better suited to times of high stress, while others may be more productive but also more vulnerable to adverse conditions.

Another team of researchers, led by Rob Rowan of the Smithsonian Tropical Research Institute, in Panama, confirmed Buddemeier's theory. They showed how, within a single reef, corals would swap algal symbionts with low light tolerance for ones with high tolerance as the levels of solar radiation penetrating the water increased, thus protecting themselves from damage.

By bleaching themselves at the onset of stress, corals are attempting to swap their crop for a more productive one, much as we might if we found a variety of a vegetable crop we had planted was unsuitable for our local climate. Rather than continuing to grow a badly yielding crop, a good gardener will always look around to see whether he or she can find a more suitable variety. The coral host closely regulates its domesticated algae by mechanisms similar to those used by the oligotrichs, and those whereby legumes regulate populations of rhizobia bacteria. There are a million green protists in every square centimeter of coral surface. By limiting the amount of nitrogen available to its garden, the coral has evolved such that these huge numbers of *Symbiodinium* do not run out of control.

An even more serious threat to corals than human-induced global warming way well turn out to be nitrate pollution from increasing overuse of artificial fertilizers in agriculture. When levels of nitrogen in the sea are high, the coral cannot stop its garden from growing. The *Symbiodinium* proliferate and the coral becomes sick. Just like the many lichens whose alliance cannot tolerate air pollution, the highly evolved gardening techniques that produce the elegant coral reef can take only so much buffeting before they become too unbalanced to be effective.

Writing in an age in which it seemed that science would soon investigate every last detail of the earth, Jules Verne could hardly have imagined how incomplete an understanding of the mysteries of the sea scientists working a century later would have. From anglerfish lures and *Ophrydium* colonies to seaweed-sucking sea slugs and crop-juggling corals, it has taken researchers until the past decade or so to uncover the many intimate associations at the heart of aquatic biology. With only a handful of exceptions, this same pattern—a long Dark Age followed by a very recent Renaissance—has been true of the study of symbioses in all the earth's major environments.

Not every beautiful feature of the marine world, however, is a product of evolution by association. Similarly, not all luminosity is created by symbiotic bacteria. Yet the ability to glow in the dark could well have originated from a bacterial liaison and then become integrated into an animal's metabolism. And symbiotic recycling of nutrients is vital for sea organisms, as surface creatures constantly lose materials to the oceanic abyss. A lack of food and light drive seafloor dwellers to ever more intricate symbiotic solutions.

It is no wonder that even Jules Verne's fertile imagination did not conjure up images as extraordinary as the products of symbiosis in the oceans. Despite Dubois's humiliating retreat in 1919, the marine realm has provided perhaps the most graphic of the recent discoveries concerning symbiosis as the source of major innovation for evolution on Earth.

CHAPTER 4

BEDBUGS and BUBBLE BOYS

You've got to eat a peck o'dirt before you die.

OLD ENGLISH ADAGE

In 1926 T. H. Morgan, a Columbia University geneticist who would later receive the Nobel Prize, announced that to understand evolution, genes were the only relevant units of study. The microscopic inhabitants of his subjects could "be ignored genetically," he proclaimed. His colleagues agreed. Yet one German biologist obstinately did not.

Paul Buchner's father was a doctor and had encouraged an interest in nature in his son. In 1907 Paul enrolled at the University of Würzburg with the intention of becoming a botanist. His attention was soon distracted, however, by the newly discovered liaison involving a yeast that lived inside sap-sucking insects. "With the publication of these studies it seemed as though a blindfold had been removed from my eyes," said

Buchner. Within three years he was working full-time on groundbreaking research into hereditary symbioses. He studied aphids and cicadas as well as ants and cockroaches. He developed new experimental techniques combining elements of microbiological and genetic technologies, which gave him a new perspective on the whole insect group. By 1921 Buchner was able to publish a comprehensive study cataloging the widespread occurrence and importance of microbial symbionts in a huge variety of insects, including blood- and sap-sucking species. By the time it was widely available in English in 1965, *Endosymbiosis of Animals with Plant Microorganisms* was an encyclopedia of almost one thousand pages. Deservedly, Buchner is recognized as the first to establish that microbes were crucial to most insects' evolutionary success.

The first insects to evolve avoided leaves altogether. They preferred the tastier fare provided by scavenging, feeding on dead organisms and the microbes associated with their decomposition. When insects finally evolved the ability to eat leaves, they did so by cheating. Like every other plant-eating animal, they entered into an association with microbes that did the hardest part for them.

Eating leaves is no easy task, as any careful observation of a munching caterpillar will reveal. Leaves are slippery, being covered with a layer of wax. Since their main function is to harvest light, they are also often in positions that are exposed to the elements. Insects get around this problem in one of three ways. They can increase their ability to hold on by using sucker legs, as caterpillars do. Alternatively, like leaf miners, stem borers, and gall formers, they can feed in places where they are protected by the plant. Leaf-cutter ants, with their huge jaws, take the problem home by slicing off sections of the leaf and transporting them back to the nest.

The dry atmosphere of a leaf's surface also presents a challenge to insects. The leaf itself has a high water content, but a few tenths of a millimeter above it there is a dramatic drop in relative humidity. In response, insects often feed in groups, eat parts of the plant where the layer of damp, stagnant air is thickest, or munch at night and rest in thin shelter by day.

There is a third and crucial hurdle for insects to overcome if they are to feed successfully on leaves—nutrition. Unlike juicy fruits and a flower's nectar, green leaves have evolved to get the most out of the sun, not to benefit animals keen to eat them. So some leaves are packed full of large, indigestible molecules such as tannins, which are part of the brown-staining, acidic residue deposited by a brew of leaf tea on the inside of a teapot. What's more, leaves lack some of the fats, amino acids, and vitamins essential for insect life. So how do insects ever gain enough food from leaves to survive?

One pioneering entomologist made films of the microbe-filled guts of wood-eating termites in the 1930s.

In 1952 Sir Vincent Wigglesworth, the eminent Cambridge University zoologist, demonstrated a dependence between onion flies and a *Bacillus* bacterium. But it was not until electron microscopy became widespread in the 1960s that most zoologists realized how widely bacterial symbionts were distributed throughout the insect world.

By the 1980s, a large proportion of insects were known to contain communities of bacteria inside their guts. Microbes were known to do everything from detoxifying chemicals and synthesizing essential amino acids to breaking down cellulose and recycling nitrogen. While genes are certainly not irrelevant to evolution, they cannot be the sole mechanism of change—bugs can clearly transform insects, too.

Today, entomologists think that virtually every insect known to science is likely to contain symbiotic microbes, usually including some in their guts. In a growing number of instances, microbial symbionts are known to have been integrated into every part of the insects' life cycle. This hall of bonding fame includes ants, aphids, cockroaches, beetles, weevils, crane flies, fruit flies, tsetse flies, plant hoppers, bedbugs, shield bugs, lice, grasshoppers, crickets, locusts, termites, and many more. In all these cases, the microbes possess metabolic capabilities absent in the sterile insect, which would usually die without their microbial alliance. Yet even today, out of two thousand or so papers concerning insects that are published every year, only a handful touch on the possible role of these associations in their biology.

Among those symbiotic studies that have been carried out, the most common subject of study has been the termite. Harvard biologist Lemuel Cleveland, one of the pioneer researchers of insect liaisons, devoted forty years of his life to the back ends of perhaps the ultimate symbiotic animal—wood-eating termites. Up to half the weight of a termite may consist of its wood-eating protist associates. These microbes contain bacteria that break down the wood's otherwise indigestible lignin and cellulose. They are passed between generations by the young termites eating the anal droppings of adult termites, which contain all the species of microbial associates necessary for the digestion process.

Many species of social termite and leaf-cutter ant use huge chambers inside their nests that are filled with crops of fungi. For the termites this is a supplemental source of nitrogen and vitamins, whereas the ants rely entirely on the fungi for their survival.

The leaf-cutter ants' gardening skills illustrate features most of which are common to both insect groups. The domesticated fungal gardens are cultivated by worker ants, which seek out leaf blades that are cut into pieces before being carried to the nest. These are then chopped into much smaller fragments of around a millimeter square. Having chewed them to a pulpy consistency, the worker adds some drops of its own ammonium-rich exudates to the mashed leaf. The compost is now complete.

A leaf-cutter ant queen starts her new nest garden by carrying fungi from the old nest. To begin the composting process, she plants a small piece of the old fungal garden in the newly manured bed. She has held this all-important inoculum in a special pouch at the back of her mouth. The queen is also the first to find some plant fragments as food for its new generation of associates. The worker ants add their defecates to the garden, just as human farmers in less industrialized agricultural systems may use their own feces to fertilize their fields. Ants actively seek out compost for their domesticated fungi, including chewed caterpillars and fallen pollen sacs from flowers. The fungal hyphae flourish, and they are harvested and used as food by the whole insect colony. Worker ants also weed the garden, removing unwanted components such as bacteria and yeast. In cases where a fungal garden has been removed from its colony, it has soon become overgrown with extraneous fungi or bacteria.

When a six-year-old nest was excavated, it was found to contain almost two thousand subterranean chambers. Each gardened chamber was about ten inches in diameter, and weighed about ten ounces. After many months, like a farmer's overcropped field, the leaf-cutter ants' garden may become exhausted. While after four years around half the chambers contain gardens, after six years this fraction drops to less than a sixth. Eventually it is replaced with a new nest, leaving the old one to decompose. In its six-year lifetime, a nest will consume around six tons of vegetation.

Ants are among the first organisms whose microbial associations have been investigated using the powerful techniques of molecular genetics. By analyzing sequences of DNA, geneticists can establish the precise hereditary patterns of the ants and their symbionts. Perhaps the most exciting use of these new genetic techniques, however, has been to explore the extraordinary horticultural skill with which ants cultivate their fungi.

Ulrich Müller, a biologist at the University of Maryland, has found that leaf-cutter ants appear to continually domesticate new varieties of fungi by taking them into their nest. Müller's team screened 862 types of nest fungi, as well as related wild fungi. DNA sequences suggest that the ants are continually refining and improving their fungal gardens. Ants even secrete their own natural antibiotics to control the growth of fungal weeds. Müller's analysis of the geographic patterns of fungal distribution suggest that the ants periodically swap crop varieties with their neighbors—even with ants from different species groups. In the same way, farmers and market gardeners in many parts of the world swap seeds with neighbors in nearby communities. Since ants are thought to have been doing these swaps for 50 million years before we started, it is hardly surprising that they are such expert farmers and crop domesticators.

Social insects such as ants and termites are perhaps the most dramatic examples of how insects can domesticate microbes. However, those of us who lived within sight of an elm during the 1970s may feel more moved by the tragically effective association between a beetle and a fungus that almost entirely removed this gracious tree from the English countryside as well as from North America.

Dutch elm disease—so named because the fungus, *Ceratocystis,* was first described in Holland—arrived in Europe during World War I. It entered the United States during the 1930s, probably via furniture imports in which the elm bark beetle had stowed away, but was subsequently eradicated. Once in a new habitat, the beetles can fly tens of miles to an elm tree, where their females lay their eggs under the bark. The beetles then bore into the wood, taking fungal spores with them. By the time the larvae hatch, the fungus has colonized the surrounding wood and forms small mushrooms on the inside of the beetle's now extensive gallery network. It is these fungal fruiting bodies on which the beetle larvae exclusively feed.

Once the larvae are fed and the associated fungus has left the tree maimed, the beetle collects the fungus into a special storage organ and moves off to another tree. Young healthy trees can be killed in as little as one or two months, while the old giants may hang on for years, the upper branches becoming stag-headed, until finally no leaves are left at all. Even if all the dead and infected wood is removed and burned, the disease can spread through self-grafting root connections between different trees. The Dutch elm epidemic was an instance where the interconnectedness of plants was exploited by an invincible invader.

In the case of Dutch elm disease, the fungal associate seems to have been domesticated by the beetle. In another case, however, the insects seem to have been domesticated by fungi. The small, dome-shaped scale insect attaches itself to leaves, twigs, or fruits of a shrub or tree. Its strawlike mouthparts suck the plant's sap. Often called an "animal lichen," the inside and outside of the scale insect become permeated with fungal hyphae. The tough fungal mat over the outside of the scale insect can be up to a millimeter thick and protects the association from attack by predators and pathogens. Even the long, needle-sharp egg-laying organ of a parasitoid wasp cannot penetrate it.

Fungal hyphae connect the outside mat to the nutrients inside the scale insect's body cavity. Although dwarfed and paralyzed, the majority of scales that are infected at any one time, and are thus predator-resistant, ensure the continued existence of a smaller fraction of nonlichenized scale insects.

A wide variety of insects take the domestication of

microbes a step further. The microbes, usually bacteria, are taken inside cells of the insect gut where, like the rhizobia that inhabit the cells of legume roots, they act as intracellular metabolic maestros. But unlike rhizobia, these bacterial inhabitants are permanent and inherited via insect eggs. The sole purpose in life of these specialized gut cells, or bacteriocytes, is to harbor these insect-domesticated bacteria.

Bacteriocytes occur in one-tenth of all insect species, including tsetse flies, carpenter ants, cockroaches, and many beetles. But the best understood example is the aphid that first inspired Paul Buchner in the 1910s. It is heavily researched partly, like leguminous plants, because it has agricultural importance. An aphid is like a living syringe that inserts its needle-shaped feeding tube into a plant's circulatory tissue and sucks up the sap. In addition to the direct damage aphids do, they carry plant viruses and other diseases, which can also reduce crop yields.

The aphids' extraordinary reproductive rate—females give birth to up to fifty offspring in a lifetime of just ten days—is due to the nutrients they can access via their bacterial symbionts. An adult weighs one-five-hundredth of a gram and yet contains around 5 million bacterial symbionts.

British symbiologist Angela Douglas, of the University of York, has carried out research on aphid and leafhopper bacteriocytes. She suggests that such specialized cells are most common where food is of particularly low nutritional value. Most insects possessing bacteriocytes consume narrow and unbalanced diets—usually the sap of plants or the blood of animals. Insects without these bacterial symbionts do not necessarily die, but to obtain the same amount of nutrition they have to ingest hundreds of times more material than if they possess the association. So the alliance allows insects to exploit food resources that would otherwise be too deficient nutritionally to be worth eating. The association seems to date more or less from the origin of the aphids themselves.

Symbiotic bacteria are often transmitted from generation to generation, not externally via feces, but by direct insertion into the base of the fertilized egg. They remain there for two days after the egg has been laid and then find their way to the offspring's gut. There, the bacteria reenter the bacteriocyte cells and carry out their task for the rest of the insect's life span.

We have already seen that in many intimate associations, such as the mycorrhizae, a thin line exists between a welcome guest and a disruptive invader. Hidden inside their cells, many insects have their own secret passenger that has dramatic effects on their biology. *Wolbachia* is a bacterial symbiont first discovered in mosquitoes in the 1930s that often changes the sex of the offspring of the organisms it infects. Use of new gene-probe techniques has allowed entomologists to identify the bug in many insects, crustaceans, mites, and even nematode worms.

Within the class Insecta, it has been found in a wide range of groups, including beetles, flies, wasps, butterflies, and moths. One survey of Neotropical insects found it in 16 percent of all those species examined. It is probably in many of the ladybugs in your garden. Having been identified as the most ubiquitous symbiont of animal cells ever described, there was concern for some years that it might even infect mammals.

Females that become infected with *Wolbachia* usually pass it on to their offspring, which they produce in greater numbers than do uninfected females. More strikingly, the normal sex determination system, quite similar to our system of X and Y chromosomes, is overridden. Individuals that are genetically male are converted into females. In some butterfly species the presence of the bacterium even seems to make females adopt the flocking behavior, lekking, that is usually performed only by males.

Most fascinating for those interested in the role of microbes in animal evolution is the recent suggestion that *Wolbachia* may have a role in generating new insect species. Parthenogenesis, for example, is a process common in insects whereby females produce offspring without intervention from males. It creates populations of asexual female clones. Cut off from normal mating opportunities, a *Wolbachia*-infected insect's characteristics may start to diverge, eventually becoming so distinct that they are unable to remate with the population from which they originated. This would, by most accepted definitions, make them a new species.

Another means by which symbiotic microbes might contribute to insect speciation concerns their effect on the newly hatched larvae of insects. Anobiid beetles feed on wood and contain pouches of yeastlike fungi in the gut. Adult beetles transmit the symbionts to their offspring by smearing the eggs with fungal cells. The newly hatching larvae eat some of the eggshell and become infected with the symbiotic fungi, which supply vitamins and essential amino acids. Without the fungal symbionts neither the larvae nor the adult beetles can grow. The possibility for speciation comes if the composition of the fungi changes, thereby changing the dietary competence of the beetle—for example enabling it to thrive on a different species of wood. The resulting strain of insect could become reproductively isolated from that from which it originated, eventually becoming a new species.

Many species of insect, such as the termites that we have already encountered, keep their domesticated microbes outside their cells, inside folds of the gut. These microbially enhanced digestive systems anticipated the greatest guts of all—the giant ruminant stomachs of the mammalian herbivores, including our own domesticated cattle and sheep. It is here that these animals' potential for undertaking multiple levels of microbial domestication can be seen most clearly.

A variety of savanna communities cover two-thirds of

Africa and, in total, about half the earth's land area. Across the continents—from Siberia to Chile—each savanna has particular characteristics. Many of the African plains are associated with an extraordinary diversity of large mammals, without which the ecosystem's function would change catastrophically. Rainfall, rather than temperature, defines the seasons on the African plains, as it drives the water evaporation cycle on which the large mammals depend. The alternate wetting and drying of the grasslands acts as a pump, renewing the mineralized nutrients via the soil's innate recycling processes, making them available for plant growth.

Large mammals seem to add to this positive feedback of nutrient replenishment. By eating and then defecating their green forage, these herbivores increase the overall rate of nutrient cycling, plant growth, and ultimately their own body weight. Surprisingly, the amount of total animal body weight per square meter is higher on the African plains than in the biological zone often perceived as being the most heavily populated with animal life—the tropical rain forest.

Like the insects, mammalian herbivores have guts teeming with microbial associates that digest tough plant material for them. Ruminants such as cattle, sheep, and deer have four-chambered stomachs that harbor complex coevolved communities that enable them to obtain their nourishment by consuming large quantities of grass.

The mouthful of grass a cow swallows first passes

through to a preliminary pair of stomachs, where it remains for several hours undergoing bacterial fermentation. The stomachs are host to a mixture of bacteria and protists. When not eating new grass, cows often ruminate, or "chew the cud." This involves a regurgitation of small quantities of ingested food, which has finished fermentation and is now softer, into the mouth for a second chew and further mixing with saliva and bacteria. The fermented grass is then swallowed for the second time, passes through a second pair of stomachs, and is then taken onward to the intestine.

The bacteria are not alone in the main fermentation stomach, the rumen. New techniques have allowed microbiologists to characterize the rumen's contents both genetically and ecologically. Each milliliter of gut contents contains an estimated half a million ciliate protists belonging to at least fifty different species. Ciliates supply their bovine hosts with at least one-fifth of their protein needs, while bacterial fermentation provides important fatty acids. Calves acquire the rumen ciliates, together with bacteria, by being licked by their mothers or by eating some of the regurgitated contents of the rumen.

The environmental conditions of the rumen are so complex that they have never been successfully duplicated, nor the domesticated microbial community grown, in the laboratory. Like the ant and termite gardens, this internal microbial compost heap needs careful tending and harvesting. The rumen walls are semipermeable, allowing chemicals to be exchanged with the blood. The rumen contents are kept at a temperature of about thirty-nine degrees Celsius and virtually free of oxygen. Just as in the bacterial communities at life's beginning, the waste products of some rumen microbes serve as food for others. Fermenting bacteria produce hydrogen and carbon dioxide as waste products that different bacteria, methanogens, use to produce methane.

So long as new grass is constantly being taken in, bacteria and protists in the cow's stomach get drawn into the same positive feedback loop in the rumen as occurs in a growing plant's rhizosphere. Digestion of the starch and proteins of bacteria by grazing ciliates releases sugars and amino acids, which the bacteria use to grow and reproduce. This creates further food for the ciliates, and so the cycle continues as long as new chewed grass is being swallowed. The ruminant also harvests vitamins and proteins from its gut garden by digesting a certain proportion of the microbial crop.

The most striking effect that ruminants have on their patchy environment is that they seem to create specialized, highly connected, miniature "grazing lawn" ecosystems. Loosely analogous to the fungal gardens of the leaf-cutter ants, these lawns occur on areas of grassland and contain plants of markedly different form and function to those of surrounding areas. By biting grass to a much shorter length than elsewhere, a particular group of ruminant grazers can increase the number of active grass shoots per unit area, leading to an increase in overall forage quality. This garden-tending behavior, uncannily similar to the way in which we tend our own lawns, has been reported in a wide range of grazers, including the hippos in the Queen Elizabeth National Park in Uganda.

In Africa, India, and many parts of Eurasia, herds of large herbivores, each with its own internal microbial garden, traverse every habitat that could support them, as they have since the first evolution of human beings. They were the early architects of the savanna landscape, with a far greater impact on the characteristics of the ecosystem than the human settlers who came onto the scene much later.

Humans also contain microbial gardens in their guts, which bear the hallmark of our association with cattle. Only human societies that have lived together with cattle for many thousands of years can digest lactose. This milk sugar is not present in human milk, but communities that have kept domesticated cattle for many generations have acquired the ability to digest it, possibly from their symbiotic gut bacteria. Conversely, human populations in China and the Americas had relatively little association with domestic cattle prior to European colonization, and most of the indigenous people of these lands therefore lack the ability to digest lactose.

In some very rare instances a human's association with its microinhabitants can go disastrously wrong. The boy in the bubble who inspired pop musician Paul Simon's song of the same name was living testimony to what happens when this delicate balancing act between man and microbe becomes fatally precarious.

On September 21, 1971, David Phillip was delivered by cesarean section into a sterile plastic chamber at Children's Hospital, in Dallas, Texas, because he suffered from X-linked severe combined immune-deficiency disease (XSCID). It had already killed his brother. Unable to produce enough white blood cells, the growing David was enclosed in larger and ever-larger sterile plastic chambers. These were the "bubbles" that kept his body free of any other living organism and also inspired his media nickname. After a controversial twelve-year life that cost over $100,000 a day to sustain, the bubble boy died as the longest surviving immune-suppressed human in history.

The bubble boy is the most famous illustration of how an organism's ability to cohabit with its microbial associates is not some peripheral or advanced capability, but central to life. Bacteria make up a tenth of our body's weight. Totaling 90 trillion cells, they outnumber our own body cells by a factor of nine to one. As well as being in our gut, they live on our skin, in our blood, and even on the transparent surface of our eyes.

Humans can achieve health only by learning to maintain their alliance with such organisms. Lacking this ability, the bubble boy was doomed from birth. No human, no organism, can exist for long as a sterile being. Human intestines contain at least five hundred identified kinds of microbes that aid our digestion, and contribute up to a fifth of our food intake.

The experimental equivalents of the bubble boy are mice that have been delivered by cesarean section and raised in germfree incubators. They develop a variety of abnormalities, including massively swollen guts and a high sensitivity to infection. It takes just a hundred pathogenic bacteria to make them sick, whereas normal mice can endure a million times as many bugs before they become ill.

For decades the biggest mystery for those studying human gut symbionts has been the question of how the helpful bacteria know where to go when they enter the gut. At least on the evidence of experiments with mice, the answer seems to be that the bacteria actively participate in the development of the organism from its birth. The molecular conversation that takes place seems very similar to the one we saw in legume root nodules and the bobtail squid. Jeff Gordon, a molecular biologist at Washington University in St. Louis, has focused on how this molecular chatter between microbes and our gut takes place.

After ten years examining the problem, Gordon's team has found that the unspecialized cells lining the small intestine are destined to become any one of four different kinds of specialized cells, depending on the kind of bacteria they come into contact with during the gut's development. Their experiment used a single gut characteristic—the presence on cell surfaces of a sugar called fucose. They then watched how it changed when they fed germfree mice different bacteria.

During the first three weeks of a mouse's development, the cells in their small intestines are coated with fucose. In germfree mice the fucose levels then fall. But if gut bacteria are added, the cells start making sugars again. To find out which particular bacterium was responsible, Lynn Bry, a member of Gordon's team, fed germfree mice one species of bacterium after another. After several negative results she added *Bacteroides thetaiotaomicron,* and the sugar production started up. She found a precise correlation between the number of bugs and the number of sugar molecules. Another team member, Lora Hooper, has found evidence that the bacteria are able to send out a signal that induces sugar production by the gut cells, but send it only when their levels of sugar are becoming low. It's a remarkably fine-tuned symbiosis that is probably happening, Gordon suggests, between thousands of different types of bacteria and our intestines.

At birth, each of us probably has a rough street map on the inside of his or her gut showing zones that are particularly suited to different types of bacteria. As bacteria arrive, their molecular conversation with the gut directs them to their correct intestinal niche. The specificity of the bacteria-gut interaction shown by Lynn Bry ensures that the useful bugs stay in position despite the huge turnover of millions of different types of bacteria that pass though the gut during our lifetimes.

What is still uncertain in all three of the best known molecular conversations between symbionts—legume, squid, and mouse—is whether gut cells send out chemicals that directly turn genes on and off in the bacteria, or whether they merely select for particular bacterial strains by judiciously secreting nutrients, displaying certain receptors on their surfaces, or using antibodies to chase off the unwanted bugs.

The increasing realization of the deep intimacy of the liaison between bugs and gut cells is being taken seriously by doctors. During a typical bout of Pasteurian paranoia in the early 1990s, the bacterium *Helicobacter pylori* became public enemy number one after it was linked to stomach ulcers. Millions of dollars were spent on the quest for new chemicals to eradicate it from human stomachs. But microbiologists have now found that it is present in around half of all healthy stomachs.

A penitent Staffan Normark at the Karolinska Institute in Stockholm suspects that it may be a normal member of the human digestive system. "I spent many years showing *Helicobacter* was a pathogen," he admits. Now he believes that like any symbiont, it can show both positive and pathogenic sides. His research team has shown that *Helicobacter* releases chemicals that are toxic to salmonella bacteria, and the pathogenic forms of the world's best known bacterium, *Escherichia coli,* such as the infamous pathogen *E. coli* 0157.

All these extraordinarily complex molecular conversations and subtle chemical interactions take place on the surface of an intestinal lining that is just one cell thick. But if our bowel is severely damaged, or if food poisoning leads to rupture of the gut wall, a life-threatening infection known since ancient times as sepsis develops. Other resident bacteria that have previously been beneficial to the body's nutrition can join the invasion and cause a systemic, life-threatening infection. And the more diverse the population of invaders, the less effective the already resistance-hampered antibiotics become.

These disastrous situations apart, humans have much to gain by respecting our five hundred or so types of bowel bacteria. The best way we can do this, according to John Cummings, a clinical scientist at the Dunn Nutrition Centre in Cambridge, England, is to increase our consumption of dietary fiber—a kind of starch that resists easy digestion. Present in beans, peas, lentils, whole-grain pasta, and cereals, fiber is healthy because it stays intact in the upper reaches of the digestive system, and arrives in the lower gut ripe for breakdown by our indigenous communities of invaluable microbial allies. The more fiber in our food, the more opportunity the bugs have to extract and recycle nutrients internally.

Importantly for industrialized humanity, with its epidemic of cancer, bacteria in the lower gut produce butyrate fatty acids that suppress the growth of cancerous cells. So a diet high in fiber helps to prevent the development of bowel cancer, which strikes about one in forty of us over our lifetimes. Aedin Cassidy, also at the Dunn Centre, found that countries with high fiber intake, such as China and India, have less than a fifth as much colon cancer as those countries whose populations are keener on processed food, such as Britain, Australia, and the United States.

Processed foods may contain chemicals that encourage the less helpful bacteria among our gut biota. Sulfates, common preservatives in processed foods, are converted by certain intestinal microbes into sulfides. Too much of these chemicals may damage the cells lining our lower gut and can lead to inflammation and ulcers. A better understanding of the role of the bugs in our gut has made nutritionists reject the old notion of dietary carbohydrate as something that is merely absorbed to provide energy. Beans and pasta ensure that your populations of "good" microbes always outnumber the "bad."

Breaking wind is widely regarded, by Britons especially, as the height of embarrassment. Yet farts are merely the sign of a happily fermenting population of bacteria. Even those of us who never eat a bean emit flatus gas around fourteen times a day. A bit of gas is surely a small price to pay for a healthy garden in our gut. Once we understand our gut's microbial friends, perhaps we will learn to love them a little more and blush a little less.

Many older Britons remember an adage from their childhood: "You've got to eat a peck o'dirt before you die." Its perhaps unintended message—that germs had a positive as well as a negative side—was right on target. During most of our evolution our immune system has been bombarded with microbes from the very moment we are born. It has learned not only to live alongside our resident bugs, but also to rely on them to maintain our sophisticated network of chemical pathways and specialized immune cells. But medical science has long dismissed this fact of life and concentrated on developing technologies that will let us kill microbes rather than understand our complex relationship with them. And now epidemiologists are beginning to see signs that our attempt to live a sterile life is provoking a backlash from our microbial associates.

An epidemic of immune-related disease is now sweeping the Western world. Asthma is the worst. A rare disease thirty years ago, asthma symptoms affect around a quarter of the populations of Britain and Australia. In the United States, asthma claims the lives of five thousand people a year. Similar trends are being reported in rates of other allergic conditions such as hay fever and eczema.

Irun Cohen, an immunologist at the Weizmann Institute in Israel, has carried out research suggesting that our hygiene fetish could even be pushing up the incidence of more serious autoimmune diseases such as juvenile-onset diabetes and rheumatoid arthritis. Her explanation lies in what has been called the hygiene hypothesis.

At first geneticists proposed, as they have for so many disorders, that there was a gene for asthma. Allergies tend to run in families, but asthma's current rate of increase is far too high for a gene to be the primary cause. The same goes for another possible culprit, a virus causing asthma-like symptoms; it simply is not present in most cases. Air pollution may increase the frequency and severity of asthma attacks by provoking an immune reaction to particles in the lungs, but an increase in particles alone does not seem to cause more cases of asthma. Along with other allergies, it is more common in relatively air pollution–free cities in southern Sweden than it is in heavily polluted cities in Poland.

According to Graham Rook, an immunologist at University College in London, myxobacteria may be what save the muckier among us from asthma, hay fever, and other allergies. Myxobacteria are a group of bacteria that live in soil, ponds, and streams, but are not part of the normal microbial community of our gut. Instead they interact with the immune system in our air and food passages. Exposure to myxobacteria early in life trains an infant's immune cells so that they are able to both fight off infections in later life and be less prone to disproportionate allergic reactions to pollen and house dust mites.

Whereas in the past, healthy drinking water contained up to a billion myxobacteria per liter, in the chlorinated water of industrialized nations the numbers are negligible. And whereas our children used to play in fields, ponds, and streams, now they are more likely to stay inside watching television or playing video games. It seems that children in urban populations of rich industrialized nations hardly come across any myxobacteria, and so we may well have shielded their immune systems from an essential learning process. As Rook points out, city people don't have gardens—either in the backyards where they live or in their guts.

The gut of a three-day-old baby born into the slums of Lahore, Pakistan, is chock-full of bacteria of all types, whereas a baby born in Sweden may have no bacteria at all, even after one week. Agnes Wold, an immunologist at Göteborg University in Sweden, who made these observations, thinks it is the Swedish baby who is, at least in this aspect of its health, deprived. The paucity of gut bugs exposes the growing child not only to asthma, Wold suggests, but also to food allergies and other immune-related diseases that are currently skyrocketing in industrialized countries.

Child health specialist John Warner of Southampton University has proposed a systematic scheme of exposing babies to bacteria of which they are deprived in the superclean homes and hospitals of the modern era. He has begun clinical trials of his technique, using normally present gut bacteria, on newborn babies. They could be either given on a spoon or squirted in by a syringe. Neonatal intensive care units already give babies lactobacillus, the bacterium found in mother's milk, to help strengthen their immune systems.

When, thirty years ago, the distinguished physician and essayist Lewis Thomas accused pharmaceutical corporations of maintaining our fear of contact with dirt to retain their market share, he must have seemed a bit paranoid. Now even mainstream magazines such as *New Scientist* are suggesting that such companies are encouraging us to use new ways of avoiding giving our immune systems a workout because, says the magazine, "it is not a moneymaking proposition." Letting your kids play in the garden doesn't increase corporate profits, whereas selling disinfectant, bleach, and medications does.

Perhaps the presence of symbionts as the cause of the "Force" in the Star Wars movies marks the beginning of a cultural shift towards treating our microbial inhabitants with more reverence. Rather than turning to pill popping and obsessive cleaning, we should make sure our diet includes the kind of foods on which our bacterial allies will thrive at the expense of their neighboring enemy bugs. Now that technology has allowed us to understand a new dimension to our bodies, we should redirect research from new antibiotics and disinfectants toward means of reducing the load of artificial chemicals and pollutants on our bodies. More of us should be enabled to escape sterile concrete jungles and to reengage instead with our microbial milieu. If current trends are not reversed future generations could look more like boys in the bubble than Luke Skywalker. It would be the inevitable price to pay for fighting, rather than managing, our microbes.

CHAPTER 5

ATOMS of REVOLUTION

A mighty space it was, with gigantic machines here and there within it, huge mounds of material and strange shelter places. And scattered about it . . . were the Martians—*dead!*—slain by the putrefactive and disease bacteria against which their systems were unprepared; slain, after all man's devices had failed, by the humblest things that God, in his wisdom, has put upon this earth.

H. G. WELLS, *THE WAR OF THE WORLDS*

Though he was occasionally criticized as a novelist, readers of science fiction love H. G. Wells. Behind his writer's mask, Wells remained the biologist to whom his teacher, T. H. Huxley (grandfather of Aldous Huxley), had given glowing tribute at his graduation in 1887, and his breadth of scientific learning gave depth to his plots rarely matched since. In one classic, *The War of the Worlds* (1898), Wells invented a realistic set of events by which aliens could first evolve on Mars, and then arrive on Earth. On Halloween in 1938, when Orson Welles read out extracts from *The*

War of the Worlds, adapted as a radio news report, tens of thousands of people thought the invasion was for real. So credible were his descriptions that they caused a panic over a substantial area of the eastern United States.

War of the Worlds echoes some of the themes touched on earlier in this book. It begins by pointing out that we could be as ignorant of life on other planets as we are of bacteria. Wells's Martians suck red blood cells out of their human prey much as *Elysia* suck chloroplasts out of seaweed. Being "absolutely without sex," his Martians reproduce by splitting themselves into two identical halves, much like bacteria. However, their downfall comes when they first come into contact with earthly bacteria. As Wells recounted, such bugs "have either never appeared on Mars or Martian sanitary science eliminated them ages ago."

Written at the end of the nineteenth century, Wells's story carries the farsighted moral that human contact with our invisible bacterial equals is essential if we are to remain healthy. A sterile world, in his imaginary universe and in ours, is a dangerous one.

The intimate associations between bacteria, animals, and plants fascinated Wells. In "The Stolen Bacillus," the first story he ever published, a terrorist gains admission to a microbiologist's laboratory and steals what he believes to be a tube of deadly germs. Intending to poison London's water supply, he breaks the tube's seal by mistake, and decides to make himself a carrier by swallowing some of the contents. But far from being fatal, the germs merely cause some blue patches under the skin. In another of his stories, a precursor to John Wyndham's *The Day of the Triffids* (1951), a plant collector buys an orchid bulb, but is overcome by its scent. Soon the flower is sucking his blood.

One of the parallels between the art of a writer and that of a scientist is the imaginative leap needed to create new ideas or theories. Few writers get very far if they merely describe exactly what they see. Similarly, the best biologists interpret their observations creatively, imagining they were the organism concerned. One microbiologist who has used his imagination with revolutionary results for evolutionary theory is Kwang Jeon of the University of Kansas. In June 1966, Jeon discovered an epidemic in his laboratory population of microbes. His inspired interpretation of the subsequent events provided the crucial link in explaining the symbiotic mechanism of evolution.

Jeon's daily research involved engineering transplants between organisms as an aid to discovering how they work. Unlike human organ transplants, Jeon's transfers were undertaken between single-celled amoebae, members of the kingdom Protista or Protoctista and usually too small to be seen with the naked eye. Unlike their even smaller cousins in the bacterial kingdom, however, each amoebal cell contains a number of tiny internal organs, or "organelles," each surrounded by a membrane. The best known of these organelles, the nucleus, stores each cell's genetic information in the form of chromosomes made up of DNA.

Over many years, Jeon's laboratory had developed a meticulous technique for removing the nucleus from one amoebal cell and replacing it with the nucleus of another. Early on his team had discovered that nuclei could be successfully transplanted among amoebae that were genetically identical. Each time, the amoeba that had received a foreign nucleus would continue to function just as well as it had before. To their surprise, Jeon and his colleagues then found that nuclear swaps between different genetic strains of amoebae could also succeed.

Amoebae, like most microscopic organisms, reproduce rapidly, given favorable conditions. At least once every other day each cell divides in two. But on one occasion, after Jeon introduced a new batch of amoebae into his laboratory to join the others he had gathered from colleagues all over the world, he was surprised to find that a few days later the latter had almost stopped reproducing. Their normally high growth rate somehow arrested, many were dying. Under the microscope, he saw that the whole batch had become infected with rod-shaped bacteria. As many as 150,000 of these bugs inhabited each amoebal cell.

In a mechanism similar to that used by some cells in our own digestive system, amoebae feed by engulfing particles of food. For an amoeba, the ingested food is usually an assortment of bacteria. In this case it seemed that the amoebae had bitten off more than they could chew. Some of the food bacteria had escaped digestion and entered the body of the amoeba, those attacked becoming the attacker. Human diseases such as typhoid and tuberculosis arise in a similar fashion when the engulfing cells of our own immune system are overcome by rapidly multiplying disease bacteria.

The infection spread to the rest of his diverse and internationally renowned collection, but having spent a large part of his career rearing the amoebae, Jeon was determined to keep the cultures going. Slowly it became clear that a small minority of infected amoebae appeared to be surviving the scourge. Jeon called his new combination of organisms "bacterized amoebae." These were fragile organisms, oversensitive to heat, cold, and starvation. After much thought, he decided to select out and nurture the fittest bacterized forms and let all the other amoebae perish.

After five years of laborious attention from Jeon and his research team, the survivors—the bacterized amoebae—seemed to regain their health. They even began to reproduce at nearly their normal rate. Jeon was surprised to find that the bacteria that had caused the illness were still living inside each amoeba. Around fifty thousand rods had become established inside each amoebal cell, but now, years later, they seemed not to be adversely affecting their host.

The real shock came when he carried out experiments that showed that not only were the bacteria no longer deleterious to their amoebal hosts, but their host amoebae could not live without them.

First, Jeon transplanted the nucleus of a bacterized amoeba into a genetically identical but bacteria-free amoeba, which came from a population sent to a colleague's laboratory before the outbreak of the original infection. He found that the previously uninfected amoeba died as soon as it received a nucleus from a bacterized amoeba. The transplants were uniformly successful, however, when he made transfers of nuclei in the opposite direction—a nucleus from an unbacterized amoeba to a bacterized amoeba. Transplants between different bacterized amoebae were also possible. It seemed that the nuclei of the bacterized amoebae could no longer exist without their bacterial associates.

To reassure himself that nuclei from bacterized cells could still be transplanted, Jeon tried to inject the unbacterized amoebae not just with the nucleus from a bacterized amoeba but with the bacteria, too. Just as he had predicted, the bacteria multiplied inside the new amoebal host, which stayed healthy.

Jeon's final set of experiments used penicillin to explore the relationships between the different elements inside his new bacterized amoebae. Penicillin kills bacteria by binding to parts of their cell walls, blocking their formation. When Jeon applied the antibiotic to bacterized amoebae, the bacteria died, followed soon after by the now debacterized amoebae. However, the penicillin had no effect on nonbacterized amoebae because they lacked bacterial cell walls to which penicillin could bind. The interdependence of the bacterial and amoebal associates had become so strong that death to one meant death to both. Invaders that were once tolerated as unwelcome intruders had become the guests, perhaps even the prisoners, of their host. When the old amoebae without bacteria were compared to the new bacterized amoebae and all the differences listed, such as growth rate, temperature tolerance, and resistance to antibiotics, the bacterized amoebae could be clearly identified as a new species.

The product of an extraordinary combination of good luck, careful observation, and ingenious experimental design, Jeon's discoveries have lent support to ideas about the role of bacteria in evolution that have been out of fashion among biologists for almost a century. Instead of bacteria always "preying on" other organisms, Jeon has demonstrated a mechanism by which they could be a driving force in evolution. The previously pathogenic bacteria had been transformed, assuming the role of an indispensable organ of the amoeba within a few years. The amoeba's nucleus had altered its status within the cell, losing some of its old functions and gaining some new ones. Relative to the 4-billion-year scale of the history of life, the genesis of a new species took only the blink of a geological eye.

Jeon's fortuitous experiments allowed some biologists to think in radically new ways. Soon after his original work he began to look at ways of applying his insights to diseases with characteristics similar to his amoebal infection. His results, and those of his colleagues, had important implications for medicine, as shown in the case of Legionnaires' disease.

An apparent consequence of the increasing use of air-conditioning systems in buildings since the 1960s, Legionnaires' disease is a modern illness. *Legionella* bacteria enter parts of the conditioning system where hot water is present and multiply rapidly. When this water is dispersed from the air-conditioning unit along with a jet of air, they remain within water droplets that may be inhaled by office workers anywhere in the vicinity. Inside the body, the *Legionella* invade white blood cells in the lung, where they reproduce, often with fatal consequences in as little as forty-eight hours. With six cases for every hundred thousand members of the U.S. population each year, the problem is not on the scale of microbe-borne diseases in developing countries, but it is has risen to prominence because it seems virtually immune from attempts to eradicate it.

Knowing that *Legionella* often becomes an internal symbiont of amoebae, Jeffrey Cirillo, a medical researcher at Stanford University, tried to adapt some of the methods described in Jeon's classic 1972 paper in the journal *Science* to find new ways of combating infection. His research team designed an experiment that compared the ability to infect humans of *Legionella* that had been harbored inside an amoeba with free-living *Legionella* that had not. They found that the amoeba not only increases the bug's virulence, but also seems to cause the *Legionella* to develop a special "coiling" entry mechanism into the white blood cells that makes our bodies extremely vulnerable. Cirillo concluded that public health authorities had been wrong to assess levels of risk of a Legionnaires' disease outbreak by the number of free-living *Legionella* in our air-conditioning systems, but should rather examine the numbers of the amoebae in the systems that harbor the coiling superbugs. The group's results also explained why mere chlorination of water was not sufficient to kill those *Legionella* because they remained basking in the lush comfort of the interior of their host amoebae.

Symbiotic associations tend to be dynamic affairs, and that between *Legionella* and its amoebal host is no exception. The bug's accommodation in the amoeba protects it from water-treatment agents, but also allows it to find refuge from a bacterial predator called *Bdellovibrio*. This residency can quickly turn into a death-row prison sentence. If the amoeba becomes stressed by a sustained drop in the temperature of the water that it inhabits, it simply digests its cargo of *Legionella* to ensure its own survival. Microbiologists are now trying to use this and other features of the symbiosis to think of new ways of combating this waterborne menace.

The *Legionella* story illustrates Jeon's lesson that there is no fixed boundary between a helper, often called a "mutualist," and an enemy, or "pathogen." The bacteria that invaded his amoebae were no more intrinsically hostile than other bugs, such as *Listeria,* that make our cheeses taste good, almost always without doing us any harm. Sometimes a host can find its cargo of bacteria simply too much trouble in conditions such as low temperature, at which point it simply eats them. In other situations deadly rivals can become indispensable to each other's survival, almost like two sworn enemies from a sinking ship having to row a lifeboat to a distant shore: they must each take one oar if either is to reach land. Yet time and again, the importance of interdependence in life's evolution has been overlooked.

In traditional Darwinism, the origin of new characteristics and species is attributed primarily to competition. T. H. Huxley, Darwin's most prominent interpreter, was particularly successful at persuading his audiences that an individualistic and competitive ethos was a driving force in evolution. Yet, as long ago as 1873, Pierre-Joseph van Beneden, the Belgian pioneer of research into fertilization in animals, suggested that a Hobbesian war of all against all was not the only mechanism of evolutionary change. He argued that social relations in natural societies were as varied as those found in human society. In both, he observed, there existed conflict, partnership, parasitism, and mutual benefit.

Even before Darwin's death, an increasing number of biologists began suggesting that permanent associations between organisms were widespread. "Parasitism," with its crude implication of disease and possible death, increasingly seemed the wrong term to describe the range of relationships between beings, which were often more subtle and complex.

The new term "symbiosis" was intended to bring together all the cases where one or more different species live on or in another. Rather than implying anything about the mutual benefit between the associates, it would, according to its originator Albert Frank, be "based on their mere coexistence." The new word was all the more appropriate since many living things that had previously been assumed to be parasites were seen, on closer inspection, to be causing little if any harm to their associate, and indeed might even confer some benefit. Van Beneden concluded that in nature there are "almost insensible gradations of differences" between parasites, mutualists, and free-living organisms.

Many intimate physical associations of organisms do lie at the opposite extreme to mutual benefit. Studies of tree-hole mosquitoes provide an example of the ability of symbionts to change their status not only from a pathogen to a helper, but also from helper to pathogen. These are the tales of the quick-change killers.

Tree holes could hardly be more different from the artificially lit, air-conditioned microbiology lab in which

Jeon's team grew their collectives. Tree holes are tiny arboreal ponds, often formed in the depression between a large branch and the trunk of a tree. To a good field biologist, tree holes can be just as useful laboratories for the study of evolution as a room of sterile test tubes and petri dishes.

Most of the time, tree holes seem rather lifeless to the naked eye. They only come to life under the microscope, where their most obvious residents are aquatic ciliate protists. Their name arises from the hairlike, waving cilia by which they propel themselves at great speed through their watery habitat. One particular ciliate finds its home in the water-filled tree holes along the west coast of North America. This habitat also serves as the primary breeding site of the western tree-hole mosquito, a flesh-piercing pest and itchy annoyance to humans. Ecologists Jan Washburn and John Anderson, from the University of California at Berkeley, spent several summers recording the remarkable changes that take place in the relationship between these two pond dwellers.

Tree holes often dry out, but when they fill up with rainwater ciliates hatch from cysts to graze on bacteria, smaller protists, and even small animals such as rotifers. Both the ciliate and its prey are eaten by aquatic mosquito larvae, which hatch from eggs once they are wetted. Having digested its meal, the larva secretes a number of chemicals. One of these compounds has a remarkable effect on the ciliates. When the chemical reaches a certain concentration in the water, these microbes transform themselves into a new form that no longer grazes in the water. Instead, the ciliate enters the gut cells of its mosquito predator and multiplies. Ciliate populations eat away at the digestive cells, often leading to the insect's death. This microscopic Jekyll and Hyde bug thus transforms itself from free-living mosquito food into a parasite, or even pathogen, of the mosquito.

The ciliate's transformation occurs within a single generation, and all members of the species exhibit the same behavior. So this is likely to be a genetically encoded response to a change in the concentration of the chemical excreted by the mosquito. This is analogous to the transformation undergone by the bacteria in our mouths when they associate into a biofilm. By contrast, Jeon's invading bacteria seem to placate their host amoebae over several generations, possibly achieving this evolution of new characteristics in the bacteria via gene-injecting sex. This might enable the bacteria to become immunologically invisible within their new hosts.

The case of the quick-change killer could have significance to the 2 million people in the nonindustrialized world who are killed each year by the mosquito-borne protist pathogen that causes malaria. Evolutionary biochemists Iain Wilson of the Medical Research Council laboratory at Mill Hill, London, and Christopher Howe at Cambridge University believe that malaria first evolved when just the kind of photosynthetic, and potentially pathogenic, ciliate that is still consumed by mosquito larvae in tree holes became a permanent symbiont of the mosquito. Their research teams have shown that there are relics of chloroplasts, still containing many of the genes for their original metabolism, inside the protist malarial agent, *Plasmodium*. After thousands of generations, this protist evolved the ability to infect not only its mosquito host, but also the mammals on which the mosquito feeds. Unfortunately, it is not yet clear how this finding could lead to a more effective treatment for this debilitating and often fatal disease. The cruel irony is that if more of the funds that have been spent on medical research into malaria over the past fifty years had been used to remove the causes of malaria—especially the pools of stagnant water that are the breeding ground for the mosquitoes—the disease would be far less common than it is with all the chemicals used to treat the parasite today.

A curiously similar example may provide us with a warning of humanity's increasingly strained relationship with the natural world. In 1996, fishermen in Chesapeake Bay, between Virginia and Maryland—an area prone to overfishing—began noticing a dramatic increase in the occurrence of ulcers and erratic swimming in the fish they pursued. Humans who ate the fish became prone to acute confusion, forgetfulness, and disorientation. Others suffered from headaches, lesions, and skin that burned on contact with water. According to Glenn Morris, a medical researcher at the University of Maryland, the prime suspect is a previously innocuous ciliate, *Pfiestera piscicida,* though other protists may also be involved. Could it be that these microbes are providing us with the sort of signal that should make us more aware of the dangers of overexploiting ecosystems?

Pfiestera and Jeon's amoebal experiments show two mechanisms whereby symbionts generate outcomes that traverse the boundary between benefit and harm. This is why arguments that equate symbiosis with mutual benefit or some kind of cozy cooperation are misleading. The mycorrhizal symbiosis shows a similar tendency. A mere change in the weather can make a fungus that had been providing additional nutrients to a plant now drain resources away instead.

Symbiosis is such a useful word because it avoids the need to determine whether just one or both of the symbionts, such as the plant or the fungus, is the net beneficiary in any given situation. Unfortunately, such nuances are lost on those who would see a term for its political, rather than scientific, value.

Consider: along with the Darwinian "struggle for existence," the catch phrase of biology in the late nineteenth century was *omnis cellula e cellula*—"everything is cellular and comes from the cell." The problem of how complex cells developed and evolved, be they single-celled ciliates or the epidermal layer of mammalian skin, became a central question of the age—and a hot button topic. Jan

Sapp, a historian at York University in Canada, suggests that the development of cell theory and social theory in the nineteenth century had a common basis. For leading theorists of the time such as German biologist Ernst Haeckel and British social Darwinist Herbert Spencer, organisms, like societies, were seen as mutually dependent communities of interdependent units. Just as individuals could not exist apart from their fellows, so cells within an organism constituted a "cell-republic" or "cell-state." Within this commonwealth, different cells carried out different tasks via the principle of division of labor, just as humans in society had complementary jobs such as that of a carpenter and a forester.

As the idea of the interdependence of cells within organisms gained adherents, it paved the way for the symbiotic revolution. The ensuing debate also foreshadowed the difficulties symbiosis would face in gaining full acceptance in the following century, despite Darwin's sympathetic speculation on symbiotic theories at the height of his fame.

In 1868 Darwin mused: "We cannot fathom the marvellous complexity of an organic being; but on the hypothesis here advanced this complexity is much increased. Each living creature must be looked at as a microcosm—a little universe, formed from a host of self-propagating organisms, inconceivably minute and as numerous as the stars in heaven." By the time Darwin died, botanists and zoologists were openly discussing this new hypothesis, which suggested that all cells, apart from bacteria themselves, might be composites of several different organisms, each carrying out different functions in a multispecies cell consortium.

Beginning in the early 1880s, symbiosis research reached international prominence for the first time. Biologists in Europe and North America agreed that all plant and animal cells might be collectives, composed of a variety of once-independent organisms. For instance, bacteria might have become transformed into the energy-providing centers of respiration in a cell, the mitochondria. Other bacteria might have evolved to become the plant cell organelles we now call chloroplasts.

By the outbreak of World War I, symbiotic associations were being studied as a subset of pathology, while disease was often studied under the conceptual umbrella of symbiosis. The continuum of different types of relationship envisaged by van Beneden was bringing coherence to a range of associations—from the pathogenic to the mutualistic—that had previously seemed poles apart. "I will try," wrote the celebrated French botanist Noël Bernard in 1909, "starting with symbiosis, to understand disease, and it is appropriate first of all to show . . . that there is no absolute distinction to be made between these two orders of phenomena."

William Bateson, one of the founders of twentieth-century genetics, recognized the importance of symbiosis as a mechanism of evolutionary innovation in his 1913 text, *Problems of Genetics:* "If we could conceive of an organism to which disease may become actually incorporated with the system of its host, so as to form a constituent of its germ cell . . . we should have something analogous to the case of a species which acquired a new factor and emits a dominant variety."

Bateson's radical suggestion was that evolution need not merely occur via the accumulation of minute genetic changes in populations of organisms, but rather that Jeon-style infections could also lead to a kind of "inheritance of acquired microbes." His thesis might, if it had been more widely discussed, have led to his being dismissed as a follower of the discredited pre-Darwinian French evolutionist Jean-Baptiste Lamarck. But instead it became buried, along with the concept of symbiosis, in one of the most pervasive periods of ideological conformity in the history of biology.

Yet the suggestion that bacteria inside cells could be anything other than disease-causing agents was initially met with some incredulity, especially among microbiologists. These influential professionals, driven by Pasteur's vision, engaged themselves in the isolation of "pathogens" in the test tube. They thus hoped to find a way of destroying microbes or rendering them harmless. To such researchers microbes were, as Jan Sapp has remarked, "no more than thieves—there to steal from the host its rightful inheritance."

Just as bacteriologists were beginning to realize the innovatory power that their minute subjects possessed, research into their associations with other organisms was all but terminated. The dawn of symbiosis research, so fruitful for biologists because it broke down barriers between disciplines, bringing together zoology and botany, pathology, and ecology, was about to be crushed. A combination of World War I and the fear of the spread of Communism after the Russian Revolution of 1917 heralded its fall from favor among biologists in the West.

Scientific thinkers of the day, such as H. G. Wells, T. H. Huxley's grandson Julian Huxley, and J. B. S. Haldane, grappled with the problem of explaining what it was in our nature that could produce warlike behavior on such a massive scale. They looked to science and in particular the now influential theory of Darwinian evolution to explain the aggression and carnage going on in their time. Many biologists settled for explanations based on an extreme and anthropocentric version of Darwin's views. So despite their enthusiasm for the symbioses they saw to be widespread in nature, Wells and Huxley stated in their 1929 textbook *The Science of Life* that the process of symbiosis was, like all biological relationships, "underlain with hostility." They even preferred to talk of the two different members of a symbiosis using the loaded terms "master" and "slave" rather than merely referring to them as associates.

The Great War was the decisive factor in the rise of "the struggle for existence" as the metaphor to explain all biological phenomena. Historian Phillip Pauly suggests that male biologists were also eager to use its rhetoric to defend themselves against taunts that their profession was being taken over by women. Especially in the United States women were by then at last becoming a significant presence in biological laboratories. Even now, there are sometimes attempts to brand symbiosis as somehow a "feminine" and therefore marginal outlook on biology.

The ethos of a universal Darwinian struggle could be popularized among scientists in a variety of ways. Book illustrators had already been influenced by Tennyson's description of nature "red in tooth and claw." Their pictures invariably featured predation as a focal point. Such bloody images of nature, whether from canvas or a slide projector, always carry a more powerful message than the results of a hundred experiments. This evolutionary vision became the just-so story of magazine articles, museum exhibits, and, later, television wildlife documentaries.

Open hostility to symbiosis also came from politically minded biologists, who equated the concept with Communism, and decreed it as being similarly subversive. Despite having been coined as a scientific term, it was condemned as being part of a dangerously motivated endeavor. Alexander Brandt, a biologist at Russia's Academy of Science, was not making matters any easier. In 1896 he wrote an essay, "Symbiosis and Mutual Aid," that overtly linked associations between organisms to a word with a radical political history—mutualism.

In Britain and France, friendly or *mutualité* societies flourished in the years after the French Revolution. At first they were formed to allow workers to deal with catastrophes, such as illness or funerals, but they also became a hotbed of socialist ideas. The most celebrated French exponent of mutual aid was Pierre-Joseph Proudhon, whose 1840 book *What Is Property?* replied with the answer "Property is theft." Proudhon's mutualist philosophy contributed, along with similar ideas such as those advanced by his ideological enemy, the Communist Karl Marx, to the socialist and anarchist movements that still survive today. He advocated the reconstruction of society as a federation of workers' cooperatives. Abhorring revolution, he believed that society could be transformed by the development of a system of mutual credit—an advanced form of bartering—through which workers could amass capital and eventually withdraw from capitalism. Proudhon's ideas alarmed the French government, which imprisoned him, and drove him into exile in Belgium. Yet, despite opposition from Marxists, his philosophy achieved widespread popularity among the poor, providing the ideological drive behind the Paris Commune of 1871.

Russian scientist, minor member of czarist royalty, and anarchist Peter Kropotkin applied Proudhon's social analysis to nonhumans. In his best-selling book published in 1902, *Mutual Aid: A Factor of Evolution,* Kropotkin drew on his studies of the vast and harsh environments of his native Russia to conclude that Darwin's theory was insufficient. As well as competition, the overemphasis on which he believed to partly originate in Britain's dog-eat-dog system of industrialized capitalism, Kropotkin believed that sociability was just as important a factor in evolution. Illustrating his thesis with examples from studies of animal behavior, he concluded that social instincts served the well-being of the species and the community.

For many evolutionary biologists, World War I was final proof that Kropotkin's theory was wrong. British and American Darwinians would accept nothing short of a declaration that evolution was indeed a war of all against all. Bracketing symbiosis and mutualism together, they dismissed them both, usually without attempting to understand either. When Paul Portier's book *Les Symbiotes* was reviewed by the English journal Nature in 1919, his careful research on symbiotic theories for the origin of animal and plant cell constituents was dismissed as idiotic. The anonymous reviewer remarked that he could only describe Portier as "a jolly good sportsman" for publishing them.

In prerevolutionary Russia, the proposal that symbiosis could be a source of evolutionary innovation gained a more serious, though still critical, hearing. During the first two decades of the twentieth century, symbiosis research was pursued by the biologists Konstantine Merezhkovskii and Andrei Famintsyn.

Merezhkovskii coined the term "symbiogenesis," meaning "the origin of organisms through combination and unification of two or many beings." In Russia, both before and after the Revolution, Merezhkovskii and Famintsyn's ideas spawned a huge research program that continues today. Yet, with a couple of notable exceptions, western European scientists have only taken symbiogenesis seriously in the past two decades. Most were misled by Kropotkin's popularization of the concept as the biological form of a political principle—mutual aid. One of the exceptions, the American biologist Ivan Wallin, remarked of symbiogenesis in 1923 that "the conception is so startling and antagonistic to our orthodox notions that without further analysis and reflection the conception appears absurd." The next sixty years were to prove how frustratingly accurate his insight was.

The "Chicago school" of theoretical biologists, whose work became prominent in the mid-twentieth-century United States, could have led to an earlier renaissance in the study of symbiosis, but again misrepresented the concept. Working at the University of Chicago during the 1930s and 1940s, Warder Allee and Alfred Emerson became convinced that they could follow in Kropotkin's footsteps by showing that evolution was not merely a war of all against all. Like both Kropotkin and the social Darwinists, they believed they could show that ethical principles had their foundations in nature. Allee and Emerson believed that there was a law of evolutionary progress in living organisms from conflict toward cooperation and toleration. And just as the cell in the body functions for the benefit of the whole organism, so the individual becomes subordinate to the larger population.

During World War II, Allee and Emerson were criticized for a theory that seemed to give a justification for Nazi-like repression—a totalitarian police state that forced self-sacrifice among individuals for the benefit of the nation. In response, Allee wrote an article for the journal *Science* that suggested a compromise between individual freedom and state control. Competition between nations, races, denominations, classes, and institutions could be beneficial if it worked for the good of the whole. Allee also suggested that recent Darwinian theories of evolution had provided a "convenient, plausible explanation and justification for all the aggressive, selfish behaviour of which man is capable." In their textbook of 1949, *Principles of Animal Ecology,* Allee and Emerson suggested that the progressive evolution of organisms was neither the fascistic one in which the group exploits the individuals, nor the laissez-faire one in which the individuals exploit the group. Rather, they expected that populations of humans and nonhumans that were most likely to survive would live by the formula "one for all and all for one" and accept that "elimination of non-conformists" could destroy the "social variability upon which progressive social evolution depends."

Symbiosis, as discussed in Allee and Emerson's now classic text, followed the flow of their wider logic, constituting "a tendency of organisms to evolve towards balanced equilibrium with toleration of other species within the community." Yet, to some extent, they saw it as a primitive and inhibiting force in the evolution of "higher social functions." Their misunderstanding of the evolutionary significance of the interdependence between, for example, autotrophic algae and heterotrophic fungi, in the lichen symbiosis, led them to misleading conclusions. Their preoccupation with the community interactions of animals was to obscure the insights of the previous century's painstaking research by the likes of Beatrix Potter, Simon Schwendener, and Anton de Bary from yet another generation of biologists. Yet attempts to practice biology without the concept of symbiosis have been like the study of disease before the discovery of infective agents.

By the time the Society for General Microbiology organized a special congress on symbiosis in 1963, there was overwhelming evidence for the ubiquity of associations between members of different biological kingdoms, the persistence of most symbioses, and their profound consequences for evolution. Yet this was the first time that botanists, zoologists, and microbiologists had come together to discuss this kind of association. Held in the wake of the Cuban missile crisis, perhaps the most dangerous point in the Cold War, the participants could not help but comment on the contrast between intimate alliances between organisms they had come to discuss, and the near genocidal hostility between superpowers. Yet unlike Wells and Huxley, they did not let global political concerns overwhelm the evidence of their own eyes. The conference was a landmark both in the reanalysis of the evolutionary implications of symbiosis and in the self-confidence of those scientists studying it. Soon a surge of interest allowed evolution by association to at last be recognized as being at the forefront of a transformation of scientific thought over the next decades. Microbes were the new building blocks of biology—the atoms in this symbiotic revolution.

If Wells, who had once reveled in the discoveries of symbiotic plant-animals and lichenized fungi, were alive today, he would surely be amazed by the length of time during which the mere mention of the word "symbiosis" by a scientist was tantamount to the "thoughtcrime" envisaged by George Orwell in his novel *1984*. With its poisoned pens, tyrannous witch-hunts, and ruined careers, the story of this simple concept has a plot that would have challenged the imagination of either author. Yet Wells might also ponder the extent to which many scientific ideas must be communicated by using metaphors that become political footballs for scientists and politicians. Even though symbiosis was created as a purely scientific term, perhaps it was naive of Albert Frank to think that he could coin a word that could exist insulated from the wider world. Despite the inevitable coevolution between new ideas and the culture that surrounds them, some scientists persist in their belief that they deal only in pure, unadulterated facts.

CHAPTER 6

REWRITING GENESIS

> I agree with Kierkegaard's assertion that the less support an idea has, the more fervently it has to be believed. Our culture is teeming with preposterous ideas, believed with unflinching faith by scientists and everyone else.
>
> LYNN MARGULIS, *SLANTED TRUTHS*

Lynn Margulis is a scientist who never read textbooks. Her professor refused to use them. Hans Ris, a microscopist at the University of Wisconsin, insisted that his students read about the great biological experiments of the past in the scientists' own words. One of the classic monographs that most gripped the young Margulis was a review by eminent American cell biologist Edmund B. Wilson of the current state of his science in the 1920s. He declared that there was "no biological question of greater moment" than how our cells had managed to have so many different things going on inside them and yet "remain an organic whole." He acknowledged that some biologists, such as Konstantine Merezhkovskii and Ivan Wallin, believed that this "organic whole" was not as whole as it might seem—that parts of the cell might have arisen via symbiosis. However, Wilson dismissed these ideas as "totally baseless" and "flights of the imagination." "To many," wrote Wilson, in a calmer voice, "such speculations may appear too fantastic for present mention in polite biological society. . . . But," he added prophetically, "they may someday call for some serious consideration."

More than "serious consideration" was given to such ideas when in 1958, to the amazement and disbelief of his colleagues, Ris announced that he, along with collaborator Walter Plaut, had found bacterial DNA inside the organelles of an alga called *Chlamydomonas*. Within a year, Sylvan Nass and Margit Nass at Stockholm University published further evidence that organelles such as mitochondria might well have originated as bacteria. Margulis was hooked. Moving to the University of California at Berkeley for her Ph.D., she wrote a thesis on the theory as applied to a protist called *Euglena,* whose chloroplasts showed many properties in line with their symbiotic origin. She became convinced that the suspicions of the pre–World War I symbiosis pioneers had been right.

There was going to be no easy victory for symbiotic theories of the origin of the cell. The magnitude of ivory tower inertia resisting any shift toward such concepts was huge. The elder statesmen of biology blocked publication of papers on the subject, and funding for research. For many years Margulis was prevented from carrying out much of her laboratory work for lack of funds. Undaunted, she responded by conducting an exhaustive search of all the published data from microbiology, cell biology, biochemistry, geochemistry, and paleontology that might support her case that most major components of a cell had a symbiotic origin. Her most exciting find was a section of remarkably sharp black-and-white film made in the 1950s.

Sitting on the doorstep of the biological station at Mountain Lake, Virginia, in 1934, Harvard parasitologist Lemuel Cleveland noticed that the local cockroaches had a curious diet. Rather than eating the usual fare of detritus, these roaches ate wood. Examining the contents of the roach's gut under a microscope, he discovered a collection of microbes so extraordinary that they might well have been from another planet. He found unique protists, some nearly half a millimeter in length—ten times their typical size—together with some much smaller protists and bacteria. They swarmed, packed together, and seemed to move to a self-generated rhythm.

By the 1940s Cleveland was renowned for his discovery that termites could live on wood only because their guts were packed with cellulose-digesting microbes. Though nearing retirement, he could not rest until he discovered the precise microbial mechanism that also allowed his cockroaches to digest this unlikely food source. He was so captivated by his microscope-aided observations that he filmed them.

What excited Cleveland was the method by which a few microbes kept themselves alive against all the odds. These protists survived by eating each other. He then made the imaginative leap for which Margulis has made him famous. The ingestion of one starved protist by another, thought Cleveland, might be the first example of a primitive form of fertilization between two organisms. Though sex was invented by bacteria, it seems to have taken the evolution of cannibalistic protists to produce what we recognize as the process of fertilization.

We are used to thinking of fertilization as taking place between a big egg and a small sperm. But in the groups of organisms that first reproduced sexually, such as the bread mold *Rhizopus* and the protist *Chlamydomonas,* fertilization takes place between two cells that look identical. The result is a fused, fertilized single cell, in the so-called diploid state.

To make *Chlamydomonas* cannibalistic, all it takes is harsh conditions such as starvation, cooling, or drying out. Margulis has recently used Cleveland's work to create an evolutionary theory in which, more than a billion years ago, similarly harsh conditions may have thus led early cells to resort to cannibalism.

Fertilization, a crucial part of the origin of sexuality, may have first evolved as an antidote to hunger or drying out. Bigger cells can survive starvation and desiccation longer. However, if cells kept on ingesting each other without digestion, doubling their number of genes and cell organs, they would become progressively more swollen until they became too big to remain functional, or else just burst. For sexual reproduction to really work, the doubled cells had to find a way of splitting again.

These ancient protists, living on windswept seashores, struggled to survive in this alternately wet and dry tidal environment in which food often became scarce. Such cyclic changes would have made it advantageous for the cells to alternate between larger doubled and smaller single "haploid" states. Double cells could survive longer during periods without water or food. On the other hand, a return to water and abundant food might favor the small haploid form once more, which would probably reproduce more quickly than its double. Any organism that evolved a way of switching between processes of cell splitting and cell fusion would therefore thrive in a changeable environment, leaving more offspring than its neighbors.

Margulis's hypothesis for the origin of sexuality is radical. She believes that the ecological relations of ancient microbes drove a process that ultimately led to our way of reproducing. She bases this ambitious idea on a theory she published in 1967. Now classic, the theory attempted to explain the biggest missing link in evolution—the jump from bacteria (often called prokaryotes), all of which lack nuclei, to modern cells, or eukaryotes, whose cells contain nuclei.

The differences between prokaryotes and eukaryotes are so profound that they make the distinction between dinosaurs and dogs or birds and bees look negligible. They are the two superkingdoms of life. Eukaryotes include animals, plants, protists, and fungi, each cell of which generally contains hundreds of times more DNA than a prokaryote.

Unlike many other transitions in evolution, there are no intermediates between eukaryotes and prokaryotes. It is as if honeybees mutated into humans without any evidence of rats, cats, or chimpanzees in between. The evolutionary processes behind this great revolution have had to be discerned without the help of one of the evolutionist's most trusted sources of evidence—the fossil record.

One and a half billion years ago, bacteria remained the only life on Earth, yet they permeated the air, the water, and the primitive soil. Teams of these all-conquering alchemists recycled gases and other compounds much as they do today. During the next half a billion years eukaryotes evolved into the nucleated cells that were to become plants, animals, and fungi. They were usually larger and more structurally complex organisms than bacteria. They had winding channels of internal membrane. One membrane enveloped a new structure, the ancestor of the nucleus in modern cells. They also had contracting fibers for pumping materials around the cell, as well as energy-producing mitochondria and sometimes photosynthesizing chloroplasts.

Rockefeller University geneticist Joshua Lederberg, with the assistance of his electron microscope, would help explain this radical change by showing that bacteria practice a sort of "gene-injecting" sex—the key factor that makes them fundamental atoms of evolution.

Animals and plants pass their genes down through the generations—a process known as vertical transmission. There can be no exchange of genetic material apart from that which occurs in the sperm and the eggs during reproduction. Bacteria, by contrast, are more likely to undergo horizontal transmission of genes within a single generation. They can transfer genes from bacterium to bacterium. The bug's outer membranes are also genetically permeable to both naked genes and viruses. Horizontal transfer processes are now known to be the major mechanism behind the spread of genes for the resistance to antibiotics among disease-causing bacteria such as tuberculosis bacteria and *Staphylococcus.*

In biology, sex refers not just to an act of union, but to the recombination of genes from more than a single source. Some organisms, like jellyfish and stick insects, can reproduce without having sex, in a process called parthenogenesis. By using horizontal gene transfer, bacteria can do the reverse; they can have sex without reproduction.

Sex conferred an important advantage on those bacteria in which it first evolved. Imagine a situation where two toxins—let's call them X and Y—are both present in an environment that is also rich in nutrients. Two sorts of bacteria are present—one able to feed on X but poisoned by Y, and the other type able to eat Y but killed by X.

So long as at least one bacterial type has the capacity for horizontal sex, it can make the best of both organisms' abilities. If the bug that eats X can also acquire genes to resist Y, then it can thrive. Similarly if the Y-resistant bacterium can gain the gene sequences that allow it to feed on X, it will also be able to grow. Either transfer would lead to an organism that would produce offspring capable of thriving in an otherwise toxic environment. No wonder these superflexible survivors turn to gene-injecting sex more often in times of stress and less often in wholly beneficial environments.

Lederberg and his colleagues reached such conclusions by studying bacteria in the laboratory. Yet studies of microbes in such artificial conditions seldom tell us much about the real world and sometimes give a misleading picture. It's like the famous story told in animal behavior classes at universities around the world about the female praying mantis, which eats its male partner after sexual intercourse. Long after the mantis's girl-power credentials had entered popular mythology, even appearing on David Attenborough's acclaimed BBC wildlife documentaries, more methodical studies showed that this was untypical behavior for the mantis. Its cannibalism was brought on by the extremes of stress that frequently exist under laboratory conditions. In a similar way, microbiologists have attempted to build a more balanced picture of bacterial sex beyond the test tube, where the bugs often live in close association with nonmicrobial organisms. Is horizontal gene transfer common, they wondered, not just in spreading antibiotic resistance in hospital wards, but also out in nature?

Microbial ecologists soon found not only that gene-injecting sex is common in natural ecosystems, but that it happens between different bacterial types so often the very idea of a bacterial species becomes suspect. If large numbers of genes—a cell's library of data—are frequently moving between different types of bacteria, then the notion of their being genetically distinct species becomes obsolete. To take a radical view, bacteria may one day be found to be so genetically cross-linked as to make them all members of a single species. This megaspecies would therefore possess just one global genome, which transferred itself fluidly between its different constituent types. Whether or not this stretches the concept of a species too far, local bacterial communities certainly employ genetic and metabolic mechanisms, almost like a collective brain, which enable them to find solutions to almost every ecological problem. Genes favorable to a particular niche are genetically shunted to strains that survive only if they receive the genetic information on time. This bacterial data network, a microbial Internet, possesses more information than the brain of any human.

Of course bacteria have no actual brain. Whereas we might choose a cookbook to decide what ingredients to buy for a meal, a bacterium can start formulating its recipe only once it has encountered the ingredients. Rather than planning for the future, they have to rely on finding a solution once the problem has arrived. Gene transfer is important for bacteria because new genes carry new recipes. The more recipes the bacteria have on their internal bookshelves, the more often they will be able to make dinner from the ingredients they encounter.

These remarkable findings not only won Lederberg the Nobel Prize, they also allowed him to rethink the previous rejection of symbiogenesis by his peers. In a 1952 paper, Lederberg pointed out the similarities between known bacterial symbionts that lived inside cells and cell organelles such as mitochondria and chloroplasts. His insight received little attention at the time. It was only fifteen years later, when Margulis published a detailed, provocative, and general hypothesis for the origin of eukaryotic cells, that the biological community began to realize the importance of Lederberg's insight.

Margulis used the clues she saw in present-day cells to project back to the world of one and a half billion years ago in which eukaryotes first evolved. A keystone alliance in the birth of this new superkingdom may have been centered on a bacterium that thrived in hot water. Its closest surviving representatives are *Thermoplasma,* which live in volcanic hot springs. In the bubbling pools of early Earth, these *Thermoplasma*-like bacteria were beset by snake-shaped bacteria that relied on their waste products. These bacterial snakes, spirochetes, were first described by the man who discovered bacteria themselves, Dutch microscopist Antonie van Leeuwenhoek. Reading his description of these "animalcules" more than three hundred years later, it is easy to see why these threadlike bacteria might have furnished the first eukaryotic cells with their ability to move:

> I have also seen a sort of animalcule that had the figure of river-eels. . . . These had a very nimble motion, and bent their bodies serpent-wise, and shot through the stuff as quick as a pike through water. . . . They moved their bodies in great bends with so swift a motion, in swimming first forwards and then backwards, and particularly rolling around on their long axis, that I could not but behold them again with great wonder and delight.

From his description, van Leeuwenhoek could have been describing something quite large, yet a thousand spirochetes could fit side by side on a pinhead. Their tiny snake-cells can attach to any surface, living or nonliving. Spirochetes also form a major component of the microbial community that Cleveland described in the int of his cockroaches. His cockroach films show spirochetes stuck to, and feeding from, the surfaces of larger microbes inside the insects' guts.

Margulis suggests that millions of years ago these spirochetes became symbionts by fixing themselves to the surface of *Thermoplasma*'s ancestors. The merger led to a composite organism that swam to new food sources easily, and left more composite offspring than those that remained unmerged. Eventually these early symbiont spirochetes structurally merged with their hosts, just as Kwang Jeon's amoebae became bacterized.

Symbiotic spirochetes, Margulis believes, became integrated into the cell before the appearance of the nucleus. Many of the moving cellular structures in eukaryotic cells subsequently evolved from these serpent-shaped, highly motile bacteria. The hairlike cilia inside your nose, lungs, and throat, and in the balance organ in your inner ear, and the tails of animal sperm: they all have the same telltale shapes and mobility that hint at a bacterial past. Perhaps most importantly, Margulis believes that her symbiotic hypothesis, if substantiated, will explain the evolution of cell division among eukaryotes.

Having neither nucleus nor chromosomes, bacteria divide by forming a bud, or they split in two, taking half the cell contents into each offspring. Nucleated cells, by contrast, divide by a complex "dance of the chromosomes," during which the hereditary material is first replicated and then moved to opposite ends of the cell after having attached to a mysterious set of filaments, spindle fibers, before the cell then divides. Such division is the means of growth for all animals, plants, fungi, and protists. Margulis's theory suggests that the spindle fibers evolved directly from spirochete bacteria by a symbiogenetic process similar to that which led to the evolution of mitochondria and chloroplasts. In pursuing her theory, Margulis was careful to avoid the mistake that had led to the ridicule of Paul Portier's book *Les Symbiotes* by *Nature,* in 1919. His radical claim that mitochondria *were* bacteria has taught his symbiologist successors to look for more subtle means to demonstrate the bacterial origin of cell components, and develop their own new terminology to describe the symbiotic phenomena they encountered.

By the time Portier's book was published, some five hundred scientific papers had focused on the origin of the mitochondria, yet confusion reigned. Part of the problem was that researchers used a range of different and sometimes conflicting language to describe the symbiotic associations they studied. Biologist Hermann Reinheimer's *Symbiogenesis: The Universal Law of Progressive Evolution,* published in 1915, introduced Merezhkovskii's terminology into English for the first time. Perhaps because it is difficult to follow, Reinheimer's book was not widely read and Portier's expression of his new theories was hampered by the germ-based terminology of the Pasteurians. While this conservative-inspired terminol- was being promoted in government-funded public across Europe, alternatives remained locked in the laboratories of a disparate band of marginalized biologists. While his opponents could draw on the influence and research funds of the Sorbonne, the Musée d'Histoire Naturelle, or the newly established Institut Pasteur, Portier's position at the lowly Institut Océanographique in Monaco did not even allow him to build his own research team.

Portier's biggest obstacle was that the Pasteurians had made a huge financial, political, and personal investment in the germ-theory terminology. The suggestion that symbiosis was a normal occurrence threatened to introduce a grand new language of biology in which their vocabulary would be found wanting. It had to be stopped. Knowing that international prestige and scientific careers were at stake, Portier condemned the conservatism of the Institut Pasteur. "Never will this establishment," wrote Portier to a friend, "equipped with the powerful means of research that you know, forgive one isolated worker for opening a way that it should have found long ago."

When *Les Symbiotes* sold out within a year of publication, Portier had no idea why the publisher refused to print more copies. Then, in November 1919, his publisher released a new title, *Les Mythes des symbiotes,* by Auguste Lumière. As brother of Louis Lumière, inventor of the first motion-picture film projector, Auguste could count on the support of one of the wealthiest men in France. He idolized Pasteur, and what he called his "immortal work," but saw it undermined by Portier's new ideas. Lumière's book not only denied the universality of symbiosis, but also denied the very existence of mitochondria in many plant and animal cells. Intended to be a wholesale refutation of Portier's claims, Lumière's book did little more than muddy the waters.

Years earlier Portier should have been awarded the Nobel Prize for his work in animal physiology, done together with Charles Richet, who instead received the prize alone. Once more on his own, Portier found it increasingly hard to swim against a Pasteurian flood that would permeate biology for the next few decades. By the time he took up a position at the Sorbonne in 1921, Portier had turned his back on bacterial symbioses. The Pasteurians had won.

In the same year, his colleague at the Sorbonne, Maurice Caullery, published a book, *Parasitism and Symbiosis,* that surveyed all known examples of the phenomena, save those that involved bacteria. It was a matter, said Caullery, "best left to the bacteriologists," by which he meant the Pasteurians. This timid tome, not Portier's masterpiece, became the standard text on symbiosis, being translated into three foreign languages. At the time Margulis first started publishing her ideas in the 1960s, as now, Portier's insights still remained unavailable in English.

Margulis's 1967 paper proposed, as Lederberg had fifteen years before, that the ancestors of mitochondria were themselves bacteria that had been incorporated into other bacterial cells. Eukaryote cells, she suggested, evolved by a process in which a prolonged symbiotic association had become permanent and heritable. Adopting Merezhkovskii's term "symbiogenesis" for the transformation, she proposed that mitochondria-like bacteria had reproduced inside their hosts without killing them, in a process similar to that which Jeon had observed in his infected amoebae. The metabolic performance of the cell consortia would have allowed them to survive in environments in which their neighbors could not. An ancestral line of symbiogenically produced cells thus became established.

The result of an interspecies merger, the new composite entity would have evolved rapidly, as had Jeon's amoebae. Margulis believes that many of a cell's components are also cryptic remnants of their past, as if cells are enthusiastic collectors of gadgets, all acquired through the process of symbiogenesis. According to symbiosis theory, each eukaryote cell is full of reproducing remnants of their past associations, which have been transformed as natural selection optimized efficiency.

The ancestor of eukaryote mitochondria, according to Margulis, may well have been a bacterial predator that was capable of both exploiting oxygen, via the respiration process, and doing without it when necessary. Microscopic predators that use this kind of metabolism still thrive today. *Daptobacter,* its name literally meaning "gnawing bacterium," burrows into its larger bacterial victim and eats its insides, dividing again and again until its prey is destroyed.

When the potential symbionts first invaded, their bacterial prey probably could not survive. But eventually, like Jeon's amoebae, some of the victims evolved tolerance to their unwelcome guests. The invaders thus remained alive and well within the food-rich interior of what was previously their victim. As they reproduced inside the invaded cells without causing harm, the former predators formed an association with this former prey, each eating the other's leftovers. The growth of each became dependent on the products of the other's metabolism. The process that Jeon observed in his laboratory could have been the mechanism that led to one of the biggest breakthroughs in the evolution of life, the ability of modern animal, plant, and fungal cells to respire—extract energy from food compounds—using oxygen.

For the next billion years, suggests Margulis, the invaded and the invader lived in dynamic ecological alliance. Eventually the onetime predators were tamed. They are now fully integrated organs of the cell—the mitochondria. Because the symbiogenesis of mitochondria was the set of evolutionary events from which most complex cells are derived, all eukaryotic organisms have remarkably similar energy metabolisms. Nearly all contain mitochondria. Humans, just like fungi, their fellow eukaryotes, depend on oxygen processed by mitochondria for energy to drive muscles or power the chemical reactions that build new tissues. Even the leaf cells of green plants use oxygen for respiration at night. Plants are overall producers of oxygen, however, because they also carry out photosynthesis during the day, which uses energy from sunlight to split water and thus make food with carbon dioxide molecules.

The discovery of DNA inside the mitochondria of eukaryotic cells in the 1970s helped to clinch the case for their bacterial origin. When this DNA was examined, its genetic fingerprint was far more similar to that of certain free-living bacteria than it was to DNA in the nucleus of the eukaryotic cell. This was telling evidence that mitochondria arose outside the host cell rather than having been synthesized on the inside. Seventy years earlier, the influential biologist August Weismann had proclaimed such a phenomenon impossible, pointing out that the nucleus had to be the sole repository of hereditary information, because of the "economy of nature." The storage of genetic information by two different parts of the cell, the nucleus and mitochondrion, could not evolve, he suggested, when a single carrier—the nucleus—would be more efficient. He reckoned without symbiogenesis.

Mitochondria have not only their own genes, but also their own reproductive timetable. They often divide out of step with the rest of the cell. No one still suggests—as did the pioneers of symbiosis theory, Paul Portier and Ivan Wallin—that cell components such as mitochondria are identical to the bacteria from which they are directly descended. But their continuing semiautonomy has led some biologists, such as David C. Smith of Oxford University, to develop imaginative metaphors for the latest symbiotic origin story.

Smith has used the cat in Lewis Carroll's *Alice's Adventures in Wonderland* to illustrate the way in which we should view present-day organisms with an eye to their symbiotic origins. In the story, Alice encounters a Cheshire cat sitting in a tree. As she watched it, "it vanished quite slowly, beginning with the tail and ending with the grin, which remained some time after the rest of it had gone." Just like the cat, suggests Smith, symbionts such as mitochondria have lost much of their original structure and appearance. They progressively lose pieces of themselves, slowly blending into the background of the cell. But like the Cheshire cat's smile, they have left behind evidence of their previous form. There are a number of strange yet important objects in a cell that are reminiscent of the grin of the Cheshire cat. For those who try to trace their origins, says Smith, "the grin is challenging and truly enigmatic."

In a film made for BBC-TV's *Horizon* series during the 1980s, *Intimate Strangers,* David Smith appears walking along the coast of one of the Channel Islands, which lie between England and France. On the sandy beach is what appears to be a layer of slimy seaweed, but is actually a key to the origin of the most productive cell component known to science—the chloroplast. As Smith begins to step on the wet sand, the green mass vanishes downward. Within seconds there is only gray-brown sand left. The green slime, Smith reveals, is a huge population of green flatworms, *Convoluta,* first described by the zoologist Frederick Keeble seventy years earlier. "Imagine," wrote Keeble, "a minute elongated fragment of a most delicate leaf, some one-eighth of an inch long by one-sixteenth of an inch broad, and you have a picture of *Convoluta.*"

But why are they green? The flatworms have merged with green algae to form composite organisms that lie in dense masses on the shoreline. These symbiotic animals do not eat. Instead they make their own food from sunlight and air, like plants. Seeing that they existed in two kingdoms at once, Keeble named them plant-animals. The algae not only live inside the tissues of the flatworm and produce food for it, but also recycle the worm's waste products, such as uric acid, and turn them back into food. The worm thus nurtures its algae with nutritious compost. For Keeble the algae thus constituted a "well-tended, highly productive little garden."

Knowing the derision in which the study of small, and therefore "lower," organisms was held by other biologists, Keeble wrote *Plant-Animals,* a book on *Convoluta*'s little gardens. He began it by remarking that "the more complex the organism, the more difficult it is to use the results of observations upon it for the purpose of generalising on important biological problems." Doubting whether "higher animals differ in any fundamental respect from the more lowly forms of life," he concluded that "the study of the lower organisms is not only to be justified but also urged on zoologists as one bound to lead to results of the greatest value." Although he may not have known it at the time, Keeble was heralding his successor's work on the origin of the chloroplast.

About 100 million years after a bacterial merger had given rise to mitochondria in the precursors of eukaryotes, a green-colored organism appeared in the cell fluid of some lineages of these microbes. Margulis suggests that the new bacterial merger arose not through infection, as was the case with the origin of mitochondria, but via ingestion, more like the way in which *Convoluta* acquires its green algae, or *Elysia* its chloroplasts. Photosynthetic blue-green bacteria were, she claims, engulfed by protists—larger microbes already capable of swimming and respiring oxygen. For these larger microbes the evolutionary motto was, therefore, "you are what you eat." Over the millennia these green bacteria were domesticated. In the meantime, the green prisoners managed to resist being digested and kept their light-trapping pigments alive. They thus became a self-sown garden.

From the origins of the eukaryotic cell onward there is a tendency in evolution for organisms to enter into associations with some of their surrounding organisms. In many cases these liaisons have resulted in one organism's being domesticated or tamed by another. This process is an organism's attempt to create a self-sustaining community in which the nutrients it needs for growth are constantly replenished. The resulting alliances between different organisms often leave more offspring than those of their neighbors who do not form such associations.

The domestication of one organism by another has been favored by natural selection on innumerable occasions, not just in the early eukaryotic cell or an obscure flatworm, but throughout the living world, with symbionts carrying out the most metabolically diverse functions, from oxidizing sulfur to generating methane. Like gene mutation, it has, as Margulis explains, its own internal logic.

> If you were a hunter-gatherer, you could do better than kill wild boar every few weeks if you could transform them into pigs, which were there all the time. That's what cells did when they went from something adversarial to some kind of co-living. Long-term association became the number one rule of the game. In symbiosis there are no intermediaries because we either have the organisms by themselves or the integrated symbiotic complex. How could you possibly have an intermediate stage?

Today, the remnants of these blue-green bacteria persist as chloroplasts, locked inside every plant. In a subtle shift of our perceptions of the plant world, Margulis suggests that plants turn toward sunlight not just because their own genes have evolved to do so, but also because they have received instructions from their internal symbionts. Without sufficient sunlight, the chloroplasts within a plant's cells suffer. Free-living bacteria, now tamed, influence plant behavior in a way few would have envisaged without the insights provided by symbiosis theory. Similarly, the Boston University ecologist Stjepko Golubic mocks the so-called higher plants, such as grasses and trees. "What most people call *higher* plants," he taunts, are "nothing but huge candelabras, used by blue-green bacteria to hold themselves nearer the sun."

In his book *The Microbial World,* published in 1963, microbiologist R. Y. Stanier made a statement that marked a change in the way we talked about evolution and laid the groundwork for symbiotic ideas: "The basic divergence in cellular structure, which separates the bacteria from all other cellular organisms, probably represents the greatest single evolutionary discontinuity to be found in the present-day world." Stanier's statement, and its gradual acceptance by the writers of school and college textbooks across the globe, began a slow linguistic transformation that would allow bacteria to be seen as agents not of death, but of dynamic change. However, this new symbiotic worldview still left one language of science in a terminal quandary. Taxonomy, the science of naming, identifying, and classifying different organisms, is now in a particular state of crisis.

Darwin drew evolution as a branching tree. In his scheme the characteristics of a population of organisms diverged from each other until the descendants were so different as to be classified as two new species. Now the neat branching of the tree of life so beloved of taxonomists is looking more like a forkful of spaghetti.

Since *The Origin of Species* it had been assumed that life had a single universal ancestor—a bacterium at the root of the tree of life from which all other organisms evolved. In the early 1980s molecular biologist Carl Woese, at the University of Illinois, began sequencing the genomes of different bacteria in order to determine the genetic code of this primordial cell. Yet by the late 1990s, though he had become the preeminent worker in the field, he decided to reject the whole branching-tree model of early evolution. He became convinced, like Margulis, that the organisms repeatedly came together and exchanged bits of themselves right from the earliest period in evolution. Different branches of the tree fuse together just like different fungal hyphae. They associate irrespective of the hierarchical boundaries assigned to them by generations of animal, plant, and microbial taxonomists.

Woese now believes that our universal common ancestor was not a single cell but more of a Noah's ark—a diverse consortium of gene-exchanging bacterial cells in loose symbiosis. As in the biblical vessel, a wide variety of different types of bacteria had a place in the ancestral microbial conglomeration. Later, with occurrences of symbiogenesis such as the evolution of the mitochondria and chloroplast, some of the thickest branches of the tree fused to create radically new units on which evolution could act.

The logical consequence of acknowledging this world of merger mania is to turn the standard classifications and naming of organisms upside down. The ubiquitous lichen is a merger of two members of utterly different kingdoms—fungi and plants. Is a lichen a plant or a fungus, when without either partner it would be nothing at all? In fact, every plant that hosts mycorrhizal fungi, and that is over 90 percent of them, is the product of the same type of evolutionary merger. Can we continue to simply call them plants without acknowledging their fungal dimension? Is a cow an animal or a microbial fermentation vessel, when without the microbes, the cow would not exist? This is an area of language where no one has easy answers. It is one problem that since the time of Simon Schwendener, taxonomists have found more convenient to ignore. On our new symbiotic planet, it could be not the associations between organisms that are the misfits, but rather the whole way humans have talked about their fellow creatures. Genesis is having to be rewritten.

CHAPTER 7

NEW GARDENERS of EDEN

Pangloss sometimes said to Candide: "All is for the best in this best of all possible worlds. . . ."

"'Tis well said," replied Candide, "but we must cultivate our gardens."

VOLTAIRE, *CANDIDE*

In 1852, seven years before he published *The Origin of Species,* Charles Darwin began a series of experiments that, "more than anything else," convinced him of the truth of his theory of evolution. He learned how to breed birds. To the gentlemen naturalists of his day this seemed a rather unsavory occupation, ill suited to a gentleman of leisure. While hunting wild animals was considered an elevated pursuit that enabled higher intellects to contemplate order and meaning in God's Creation, rearing pigeons and poultry appeared not only plebeian, but also unlikely to lead to any significant new discoveries. Yet Darwin hoped that domesticated plants and animals might hold the key to understanding evolutionary change.

A recent subscriber to the *Poultry Chronicle,* Darwin began to amass a collection of some ninety birds of sixteen different types. The names of the pigeon varieties give a hint of their diversity: there were pouters and old fantails, laughers and scanderoons, the snow-white and the speckled brown. He fraternized with pigeon fanciers in the beer halls of Spitalfield, in the heart of Victorian London. Flat-capped workingmen gave him priceless information as to how they picked their favorite bird and mated it with another, also chosen for certain attributes. Though far from our image of laboratory biologists, these fanciers had a sophisticated understanding of their birds. Like modern breeders, they worked with minute variations between pigeons—the very stuff of evolutionary change.

It took months of tutoring in the capital's alehouses before Darwin could spot the subtle differences on which his tavern teachers, and the judges of pigeon competitions, made their selections. Then he was ready to begin watching the process of what he later called "artificial selection" in action. The weak, ill-adapted birds would be discarded, while the good ones formed the new breeding stock. Over successive generations particular trends were encouraged and eventually these characteristics became the norm. Darwin could see the parallels between the pedigrees of the birds as charted by the pigeon fanciers and the evolutionary connections between wild organisms in general. When, in *The Origin of Species,* Darwin talked of the "hidden hand" of selection, he was thinking back to the pigeon breeder selecting his favorite bird.

In his own breeding experiments, Darwin observed the mixture of inheritance and variation the pigeons displayed. He looked at hundreds of live birds, killed them at various stages from embryo to adult, boiled up their bones, and stuffed their skins. He examined the differences in blood corpuscles between breeds. He was careful to develop what Barbara McClintock, the Nobel Prize–winning geneticist, has called, in another context, "a feeling for the organism."

Darwin also compared humanity's domestication of poultry with that they achieved with dogs, cattle, and cultivated plants. He looked back at his notes on the relationship between the "savages," as he called them, which he had encountered on his *Beagle* voyage, their animals, and their environment. He looked at the process in the round, in the widest possible context. Only then did he bring together a synthetic and comprehensive theory, his "principle of artificial selection," the description of which took up the first chapter of *The Origin of Species.*

Without the insights of Darwin's theory of evolution we could not begin to comprehend the origins of life's improbable associations. None of the diverse liaisons of life—from termites to toadstools, flashlight fish to fir trees—contradict the central tenets of evolution. Yet the fundamental importance of biological associations points to the need for a new integrative field of study, which might be called ecological Darwinism. It brings into sharper focus three of evolution's core processes: innovation, interdependence, and dynamism. One of the most, if not the most, important *innovative* elements in evolution are microbes; the *interdependence* among organisms is at least as important as the competition between them; and all intimate alliances such as symbioses are inherently *dynamic* and involve a high degree of plasticity on the part of at least one of the associates.

In every evolutionary event in every taxonomic group we have examined, from the oceanic abyss to the Arctic Circle, microbes of all kinds—bacteria, protists, and fungi—have been life's primary innovators. It is by forming liaisons with these microbes that larger organisms were able to acquire new metabolic abilities and thus occupy a diverse range of challenging niches. Symbioses have been the origin of most of the major events in the evolution of life—the transition from prokaryotic to eukaryotic cells, from seaweeds to land plants, from scavenging to plant-eating insects and the invention of sperm-egg fertilization in animals. More controversially, microbial alliances may have even been the innovative force behind the evolution of the immune system.

The question for biologists is this: what proportion of all evolutionary events over the billions of years since life began have been symbiogenic events?

During the 1990s this question was highlighted by a controversy touched on in the last chapter: whether movement in eukaryotic cells arose from a symbiogenic event involving the snake-shaped spirochete bacteria. The idea was most recently revived by experiments carried out by a team of cell biologists at Rockefeller University, led by John Hall and David Luck. They found what they identified as DNA in the part of a eukaryotic cell's internal transport apparatus called the centriole. This structure has a key role in the movement of a cell's chromosomes at the time of cell division. Curiously, centrioles seem to be able to replicate independently of the nucleus. No other parts of the cell can do this, apart from those elements such as mitochondria and chloroplasts that are now acknowledged to have arisen from symbiotic bacteria.

Lynn Margulis saw Hall's findings as vindicating her belief that symbiogenesis plays a key role in the origin of three classes of hereditary cell components. She congratulated him on what she called his "elegant experiment." However, two other research teams used what they claimed were more sophisticated genetic techniques and subsequently published reports stating that they could find none of the DNA reported by Hall. Yet neither challenger had repeated Hall's original experiment, and the Rockefeller team stood by their original conclusions. Though tantalizingly close to a resolution, the issue remains one of the controversies that surround the major events in the evolution of life.

When the first symbioses, such as Beatrix Potter's lichens, came to light, biologists assumed that such liaisons were unusual and unimportant wild cards—atypical evolutionary processes. Now it appears just as reasonable to ask why symbiosis with microbes appears to play little part in innovations in our own group, the mammals.

Despite the mitochondria in each of our body's 10 trillion (10^{13}) cells and the billions of symbiotic bacteria in our gut, we mammals have explored our symbiotic potential relatively meagerly. Why have humans become one of the most numerous organisms on Earth without undergoing further symbiogenesis? Why have we not followed *Elysia* the sea slug and put green photosynthesizers on our backs to make our food, becoming *Homo photosyntheticus?* Why do we excrete urine, a liquid containing chemicals rich in nutrients, when we could tap it to an internal microbial symbiont and reap the rewards? As yet the answers to these questions remain speculative. Only if the current wave of interest in symbiosis persists will we be likely to find the answers.

The ecological Darwinist tenet that suggests interdependence in evolution has been at least as important as competition might appear to be stating the obvious. But many of the most influential evolutionary biologists of the past fifty years have focused almost exclusively on competition. My favorite recent case is that of an article in *Trends in Ecology and Evolution,* a journal that is widely read in colleges and universities around the world. Its authors suggest that a seed lying on the forest floor is a likely site of "inter-kingdom" competition between a herbivorous animal and soil microbes. Yet in reality, the animal could digest the seed only because of the microbes in its gut. The microbes in turn are far more reproductively successful if the animal gets to the seed first, as then the microbes can feed on the chewed and digested leftovers in the animal's droppings, rather than be stuck on the outside of a tough seed coat. Where the authors of the article emphasize competition, we can now see the metabolic potential for at least as much interdependence.

Darwin was keenly aware of the interdependence between organisms. For him it was exemplified by the process of domestication. In *The Origin of Species* he uses the words "domestication" and "domestic" more than twice as often as he does the words "compete" and "competition." And although Darwin repeatedly referred to organisms' "struggle for existence," within the word "struggle" he intended to encompass "several ideas . . . [including] the dependency of one organic being on another." Darwin scholars now doubt that the supposed champion of the progressive power of competition did in fact believe that competitive forces were of supreme importance in evolution. Cultural historian Gillian Beer of Cambridge University has contrasted Darwin's "unwillingness to give dominance to a militant or combative order of nature" with the later "red in tooth and claw" interpretations put on his work.

Darwin's ideas about domestication help contemporary biologists interested in symbiosis to develop a new theoretical framework that draws on his idea of artificial selection. Imagine you are a symbiotic plant attempting to influence your surrounding organisms in a way analogous to Darwin's breeding of his pigeons. The relationship between your roots and their sphere of underground influence—the rhizosphere—will play a key role in your health and growth. One innovation you might consider would be to nurture those microbes in your rhizosphere that are most nutritionally supportive. Another might be to inhibit infectious pathogens or other organisms that interfere with these tamed and useful microbes. As with Darwin and his pigeons, you are facilitating the breeding of favored types, while suppressing the reproduction of rejects—a kind of biodomestication.

Using techniques from a new discipline, molecular ecology, geneticists have recently found that this kind of domestication process is exactly what plant roots seem to achieve with their surrounding microbes. These relationships have arisen through natural selection, the same mechanism described in Richard Dawkins's classic book, *The Blind Watchmaker* (1986). Just as fungi domesticated algae in the evolution of lichens, plants have developed a well-tended garden in their rhizospheres. There has not, of course, been any grand designer at work, just the forces of evolution—the appearance in a population of new heritable variation, followed by natural selection via the blind watchmaker of natural selection. The revolutionary ability of plants to domesticate some of their surrounding bacteria is perhaps best illustrated by genetic studies of the legume symbiosis.

Plants in the legume family such as peas, lupines, and soybeans form a unique symbiosis with rhizobia bacteria. Nitrogen-fixing root nodules are the result. Each variety of legume tends to be compatible with only a single type of rhizobial symbiont. Experiments attempting to cross-inoculate one sort of legume with rhizobia that normally associates with another type rarely produce working symbiotic root nodules. Since the 1960s, it has been known that compatible types of symbiotic rhizobia tend to be found in the rhizospheres of their host plant, rather than those of other plants. Pea rhizobia are found in the rhizospheres of pea plants, soya rhizobia in those of soybean plants, and so on. Furthermore, legume roots seem to release chemicals that are particularly beneficial to their specific symbionts.

At first it seemed that this specificity could easily be explained in traditional competitive terms. The "right" bacteria were obviously better at surviving in a particular plant's rhizosphere than their noncompatible competitors. Each was better adapted, it was said, to its particular niche. But there is a problem: the possibility of "cheats." These are organisms that could evolve to exploit the nutritionally enriched rhizosphere habitat without undergoing a reciprocal symbiosis with the plant. They would reap the benefits of the plant's nutrients without giving in return the fruits of nitrogen fixation. If such cheats became common enough, the whole symbiosis between legumes and rhizobia would break down.

Anyone who digs up a legume, be it a row of garden beans or a clump of wild clover, will soon see that such cheats never do gain the upper hand. The symbiosis is remarkably stable, and has been so for millions of years. The solution to the paradox came in the form of a gene found in the rhizobia. Studying the rhizobia that nodulate pea plants at Norwich's John Innes Centre, microbiologist Andrew Johnston found that the bacterial genes that enable the rhizobia to exploit the tailor-made food exuded by the legume roots are physically linked—on the same ring of DNA, called a plasmid—to the genes that make the rhizobia undergo successful symbiosis. So if rhizobia want to eat their special food supplement provided by the plants—the chemical homoserine—they cannot be cheats. Each legume-rhizobia association seems to have thus evolved a means of allowing the plant to cultivate a genetically domesticated strain of rhizobia, while keeping the cheats at bay.

Rather than pigeon breeding, a better analogy to describe the legume rhizosphere is that of a garden in which the valued crops are fertilized and their weedy neighbors are kept at bay. One prediction from this theory is that rhizobia with this symbiotic plasmid should turn up more often in the rhizosphere than in the rest of the soil. Molecular ecological studies with pea plants have indicated that such is indeed the case. Surveying all known symbiotic relationships involving microbes, from Arctic lichens to tropical corals, Angela Douglas, a symbiologist at the University of York, has found that the symbiotic form of a microbe is rarely if ever found far from its gardener. In just the same way, you would be very surprised to see a patch of your garden carrots thriving in the middle of a neighboring natural meadow. Cultivated crops rarely survive outside a farmer's field or gardener's vegetable patch.

Legumes are not the only plants that domesticate the bacteria around their roots, and weed out invaders. Microbiologists Patrick Mavingui and Gisèle Laguerre of the University of Nancy, in France, studied the bacterial communities surrounding the roots of wheat plants. The genetic fingerprints of these microbes suggest that the bacteria around the root all come from one genetic type, whereas those in the surrounding soil come from at least three different groups. They found that the wheat secretes a special chemical, sorbitol, a sugarlike substance we use as an artificial sweetener. The plant's roots seem to use the sorbitol to select the most useful bacteria from the soil, but the mechanism is still unknown.

Not all domestication leads to legume-style symbiotic integration. The domesticated rhizosphere of the wheat plant has remained a loose association, and no permanent merger of organisms has resulted—at least not yet. What is clear, however, is that domestication arrangements between organisms and their microbial associates, even just in the realms of life explored in this chapter, are not random, and may be more the rule than the exception.

A few years ago, having carried out experiments demonstrating biodomestication in the roots of legumes, I became interested in trying to model what was going on in evolutionary terms. At first it seemed like an impossible task. When I showed the number of different organisms, genes, and feedback loops involved in the legume-rhizobia symbiosis to several modelers, most thought it impossible. Stephan Harding, based at Schumacher College, in Devon, England, did not. He took up the challenge of helping me model the mechanism whereby legumes domesticated rhizobial bacteria.

To be successful, we had to show that our proposed genetic mechanism provided an evolutionarily stable strategy. The model consists of a series of equations with seven different variables (that's *x*s, *y*s, and *z*s) and twenty-five different feedbacks. Most biological modelers deal with only a couple of variables and one or two feedbacks. This model was bigger than anything either of us had seen before.

Having devised some equations with the help of Stephan's modeling software, we sat patiently while his PC explored the evolutionary stability of legume biodomestication. After a few simulated seasons of symbiotic interaction a pattern emerged. Domestication was there, but it happened in a pulse, when nitrogen levels in the soil became low. As levels of nitrogen fixed by the symbiosis increased, so the need for symbiosis decreased. But after the breakdown of the symbiosis, the plant would gradually drain nitrogen from the soil, until levels became so low that conditions favored the domestication of a new set of rhizobia. We tried running the model for a hundred, then a thousand seasons. Rather than fading out, or running away into unsustainably indefinite positive feedback, as we had feared, the domesticated association had formed a strategy that appeared evolutionarily stable.

The liaisons of life are nothing if not dynamic associations that are prone to change, depending on the environmental conditions. As in many marriages, the costs and the benefits on each side may alter, depending on the context within which they occur.

The preceding chapters have sampled only a tiny fraction of the millions of symbioses between microbes and larger organisms that have existed over vast tracts of geological time. Again and again, symbiosis seems to have evolved into some kind of domestication arrangement, where one associate in the symbiosis seems to have evolved to cultivate its particular microbe; not only legumes with their domesticated nitrogen-fixers, but *Kentrophoros,* the ribbon worm, whose coat of geometrically aligned bacteria provide it with a perpetual picnic. Then there are the orchids that carefully domesticate their fungi to supply nutrients to their roots without invading the rest of them. Similarly our gut bacteria help us digest food without themselves digesting us. But in both orchid and human cases, the domestication is not complete. There are cases of malnourished orchids being consumed by their cultivated fungus, while in humans damage to our gut lining can give our associates a free route to infect other organs.

Corals have evolved an advanced domestication system in which they can swap the microbes they cultivate, depending on the environmental conditions they encounter. Though they are, strictly speaking, closer to a system of full-scale biohorticulture than a permanently attached symbiosis, the masters of domestication must be the termites and leaf-cutter ants. They have devised the ultimate system of fertilizing, weeding, and even breeding their microbial crops to an extent that makes our own modern methods of industrial agriculture look biologically primitive.

The ecological version of Darwinism outlined here integrates organisms and their genes with their surroundings. By showing that most organisms survive only by the constant management of their relations with a ubiquitous part of their environment—the microbes—this book aims to contribute pieces of a jigsaw, the gradual completion of which can take biology beyond an overemphasis on the genetic code. Computer models have merely added their persuasive powers to an argument many soil ecologists have been making for years.

Any organism studied in abstraction from its interdependence with its surroundings is an unsustainable and ultimately lifeless unit. Harvard evolutionary biologist Richard Lewontin has described the essence of life as not a double, but rather a triple, helix, with the third strand representing this interdependent context, present in the life of all organisms. Domestication is a powerful concept in this respect because as one organism alters the environmental context of another, the two associates together may become a new interdependent unit on which natural selection can act. Once domesticated, microbes occupy a fresh niche in which they provide novel ways of obtaining resources that are then used by themselves and their domesticator. Together, the two organisms comprise an emergent individual whose origin and essence is interdependence.

So the first two key tenets of ecological Darwinism seem robust: that microbes are one of the most, if not *the* most, important innovative factors in evolution; and the interdependence among organisms is at least as important as competition between them. The third, which suggests that intimate associations such as symbioses are inherently dynamic, holds perhaps the most radical implications for evolutionary theory. Today's mutualist can become tomorrow's parasite, depending not on what genes it contains, but on what fortunes the fluctuation in its environment may bring. A fungal mycorrhiza can be a net drain or a net contributor to a plant's nutrition, depending on a variety of factors such as soil type, season, and even the weather. The widespread bacterial symbiont of insects, *Wolbachia,* brings advantages but also fatalities among a wide variety of insect species. Some of the bug's effects differ, depending on what point in their life cycle the insects have reached.

In 1904, the chestnut blight fungus was imported into the United States and started to infect chestnut trees, apparently causing them to die. Yet today these trees are still alive. Instead of dying, they have regrown as bushes of seven or eight feet in height. If any shoots get above this height, the fungus develops and they die back. The bush now occupies a new niche in American forests, and the fungus even seems to give the tree some protection against viruses. So is the fungus a parasite or a mutualist?

Biological alliances are inherently dynamic phenomena. Rather than discrete categories, the terms "mutualist," "parasite," and "pathogen" are better seen as fuzzy points on a continuum, along the length of which an association between two organisms may fluctuate. For many associations, the point they occupy on this continuum is as difficult to assess as it is to say who gains more on average in a marriage between two human partners. Describing symbiosis in their 1929 textbook, H. G. Wells and Julian Huxley showed that they understood this variability well:

"As in the more plastic of human relationships, casual association may pass over into mutually helpful partnership, or transform partnership into parasitism. How difficult it may be to distinguish between service and slavery."

Biologists have even managed to engineer a move from pathology to mutualism in an association between a watermelon and a fungus. Two plant geneticists, Stanley Freeman and Rusty Rodriguez, of the University of California at Riverside, transformed a fungus that caused disease in watermelons into an internal associate that actually seemed to enhance the melon's growth. They achieved this turnaround by changing a single gene.

The idea of a dynamic continuum of associations also applies to how intimate they are. At one end of the continuum is the mitochondrion or the chloroplast in which a bacterium has become so integrated into the metabolism of another cell that it has undergone a permanent merger, losing its independent existence and many of its genes. These intracellular symbioses, where the bacterial descendants have become permanent inhabitants of the cell, are inherited through the sex cells and are automatically present in every cell of the offspring. At the other extreme are the termites and leaf-cutter ants. Their cultivated fungi farms are passed on to future generations of termites and ants, but only by a process of reinfection, as the young acquire the fungus from the feces of adults. It is a cyclic rather than a permanent association. By comparison, the legume-rhizobia symbiosis is a mixture of the two extremes: the relationship is intracellular, but has to be re-formed yet again every summer as pea and bean seeds germinate and develop roots.

The ideas and theories described in this book would be of limited interest if they were relevant only to purely scientific squabbles about how life began, innovated, or evolved. Yet an understanding of the liaisons of life can, as we have seen, do far more. An ecological Darwinist approach has vital implications for the way biological theories are applied in the real world, whether through medical, agricultural, or other biotechnologies. An evolutionary perspective not only points to possible solutions to our current dilemmas, but also helps explain how we went wrong in the past. Our cultivation of legumes, the plant family that includes many of the world's staple food crops, provides a particularly striking example of how scientists have neglected the key features of intimate biological associations.

In the mid 1980s rhizobia were among the most highly researched nonpathogenic microbes on Earth. Using recombinant DNA and other genetic engineering techniques, geneticists promised strains of the bacteria that would outcompete any indigenous bacterium and double farmers' crop yields. After ten years of research, not a single genetically engineered rhizobial product had been successfully marketed, and biotech companies had withdrawn from the research. Despite thousands of field tests aimed at determining the best rhizobia strain for a particular crop, the worldwide market for inoculants remains insignificant, because farmers find that their yields do not improve. Scientists discern a small improvement in yield using inoculations of the engineered strain in a laboratory greenhouse, but in field conditions the beneficial effect disappears. What went wrong? One factor seems to be that legumes prefer bonding with rhizobia that they have domesticated themselves, rather than with rhizobia that scientists had sought to genetically engineer for them.

There is another evolutionary reason for the failure of genetically engineered symbiotic bacteria. Unless a permanent merger has been formed, such as a mitochondrion or a chloroplast, it would be in no organism's evolutionary advantage to commit itself to just one breed of domesticated associate. Instead, many organisms put their symbionts out to tender with every new generation, rather than rely on a monoculture of any single variety. The algae-juggling corals and fungi-swapping ant and termite colonies are illustrations of this principle. Similarly, preindustrial agriculturists, modern organic systems, and most of today's Third World farmers try to keep a range of varieties of any given domesticated crop on their farms, in case environmental conditions demand one rather than the other. Another reason monocultures are rare in nature is because they make life easy for pests. Genetic uniformity in an agricultural system is analogous to providing every house in the neighborhood with the same lock. If burglars can get into one home, they can get into any of them.

In the rush to identify genes for drought resistance and salinity tolerance, commercial biotechnologists may ignore the microbial alliances that allow many plant species and traditional crop varieties to survive such harsh conditions. For centuries a wide range of crops have been given this resistance by naturally occurring fungal symbionts. These integral associations are the first casualties when genetically engineered varieties, and their attendant fertilizers, herbicides, and pesticides, replace complex, cultivated biodiversity. As Kansas farmer and philosopher Wendell Berry has said, these people do not know what they are doing "because they have no idea what they are undoing." The fields where the companies grow their crops are so permeated with chemicals and lacking in natural compost that the indigenous fungi and other microbes find it hard to survive.

Yet a new respect for the self-regenerative dimensions of cultivated systems is growing. Organic or "natural systems" agriculture draws on the wisdom of farmers who, over thousands of years, have learned to work within their environmental constraints. These farmers constantly strove to breed a diverse range of plants and animals that could ensure a crop despite variations in climate, pests, and soil conditions. Such methods arise from a complex ecological understanding of, combined with sympathy for, the land and its organisms. Various research teams, including one led by David Pimentel at Cornell University, have shown that well-managed organic systems can produce yields just as high as chemical-based agriculture.

At a recent exhibition held by an association of agrochemical companies in Lithuania, the country's deputy prime minister received them with the following greeting: "You are very welcome to come and meet us here in Lithuania, but your style of farming has no future in this country." Fields in the former Yugoslavia, in which land mines kept farmers from growing fertilizer- and pesticide-sprayed crops for several years, now give them the perfect ecological conditions to return to the organic methods of their ancestors. Farmers across eastern Europe are resisting the overtures from chemical companies and are instead using the organic techniques passed down from a previous era.

In India, which is one of the prime targets for the transnational biotechnology companies, my research in some of the poorest regions over the past few years has convinced me that genetically engineered crops come very low on the list of priorities. When asked what agricultural improvements affect their livelihoods most significantly, most marginal farmers ask for easier access to water, and farmyard manure to restore fertility to land robbed of organic matter by decades of use of government-promoted chemical fertilizers.

After two hundred years, all the while neglectful of the

extent to which our activities are ultimately constrained by our ecological relations, humanity may, like Voltaire's Candide, at last be becoming more conscious of its green-fingered essence. An approach based on nurturing our living liaisons could bring about a revolution, not only in how we produce our food, but in what we think life is, and how we have evolved.

Theodor Dozhansky, a Russian émigré and the father of the modern evolutionary synthesis, famously said that "nothing in biology makes sense except in the light of evolution." If we accept an ecological Darwinist approach, nothing in the living world will look quite the same again. Perhaps its most urgent practical message is that we need to become more actively aware of the full extent to which our futures, like those of all organisms, are ecologically bound up with those of a larger biological whole. Whether by destroying our soils with industrial agriculture or polluting the atmosphere with our fossil fuels, we risk disturbing this living global system beyond its capacity to support us. Our planet could move to a state that is no longer as hospitable. Perhaps we are already reaching that point. As products of 4 billion years of evolution by association, we need to learn rapidly to be wise gardeners of our Eden, or we risk having the worst of all possible worlds.

Further Reading

Additional information and background material are available via the Internet at www.wiley.com.

Introduction

Dixon, Bernard. *Power Unseen.* New York: W. H. Freeman, 1996.

Docherty, Michael, and Stephen Tomkins. *Brine Shrimp Ecology: A Classroom-Based Introduction to the Learning of Ecology.* London: British Ecological Society, 2000.

Geison, Gerald L. *The Private Science of Louis Pasteur.* Princeton, N.J.: Princeton University Press, 1995.

Chapter 1
Beatrix versus the Botanists

Douglas, Angela. *Symbiotic Associations.* Oxford: Oxford University Press, 1994.

Findlay, Walter Philip Kennedy. *Wayside and Woodland Fungi*. London: Warne, 1967.

Holdredge, Craig. *A Question of Genes.* Edinburgh: Floris Books, 1996.

Chapter 2
The Wood Wide Web

Kendrick, Brice. *The Fifth Kingdom*. Newburyport, Mass.: Focus Publishing, 1992.

McMenamin, Mark A. S., and Dianna L. S. McMenamin. *Hypersea: Life on Land*. New York: Columbia University Press, 1997.

Woodham-Smith, Cecil. *The Great Hunger: Ireland, 1845–1849*. London: Penguin, 1991.

Chapter 3
Hidden Gardens of Atlantis

Herring, Peter J., ed. *Light and Life in the Sea*. Cambridge, Engl.: Cambridge University Press, 1980.

McMenamin, Mark A. S. *The Garden of Ediacara*. New York: Columbia University Press, 1997.

Smith, David C., and Angela Douglas. *The Biology of Symbiosis.* London: Edward Arnold, 1987.

Chapter 4
Bedbugs and Bubble Boys

Jablonka, Eva, and Marion J. Lamb. *Epigenetic Inheritance and Evolution: The Lamarckian Dimension.* Oxford: Oxford University Press, 1995.

Lewontin, Richard C. *The Triple Helix: Gene, Organism, and Environment.* Cambridge, Mass.: Harvard University Press, 2000.

Piel, Gerard, and Osborn Segerberg, eds. *The World of René Dubos.* New York: Henry Holt, 1990.

Chapter 5
Atoms of Revolution

Margulis, Lynn. *Symbiosis as a Source of Evolutionary Innovation.* Boston: MIT Press, 1991.

Reed, John R. *The Natural History of H. G. Wells.* Athens: Ohio University Press, 1982.

Sapp, Jan. *Evolution by Association: A History of Symbiosis.* New York: Oxford University Press, 1994.

Chapter 6
Rewriting Genesis

Dyer, Betsy Dexter, and Robert Alan Obar. *Tracing the History of Eukaryotic Cells.* New York: Columbia University Press, 1994.

Gould, Stephen J. *Wonderful Life: The Burgess Shale and the Nature of History.* London: Hutchinson Radius, 1989.

Margulis, Lynn. *Symbiosis in Cell Evolution.* 2nd ed. New York: W. H. Freeman, 1993.

Chapter 7
New Gardeners of Eden

Baumann, Martin, ed. *The Life Industry: Power, People, and Profits.* London: Intermediate Technology Press, 1996.

Darwin, Charles. *The Variation of Animals and Plants under Domestication.* Baltimore: Johns Hopkins University Press, 1998.

Pretty, Jules N. *Regenerating Agriculture: Policies and Practice for Sustainability and Self-Reliance.* London: Earthscan, 1995.

Acknowledgments

Many friends and colleagues have contributed to the ideas developed in this book. I am grateful to Max Walters, Robert Sternberg, and Nicola Baird for invaluable advice on the very first draft and to Sarah Corbet, Lynn Margulis, and Sir David Smith, who have generously spent time making extensive comments on my evolutionary arguments. Alastair Fitter, Jim Harris, Caspar Henderson, and Stephen Tomkins kindly scrutinized some or all of the final manuscript. I much appreciate their help. My thanks also to my editor at John Wiley, Stephen S. Power, for his diligence and support. The responsibility for all errors and inconsistencies remains, of course, my own.

More than anyone else, it is my tutors who fired my interest in biology. At Hills Road Sixth Form College, Cambridge, Phillip Nicholson and biology teachers Peter

Bilton, Patrick Holden, and especially Stephen Tomkins and Michael Reiss exuded infectious enthusiasm, kindness, and patience. At Cambridge University, Janet Moore's bubbly and exciting tutorials were the highlight. I valued the open, yet critical intellects of teachers such as Patrick Bateson, Henry Disney, Bryan Grenfell, Peter Grubb, Harmke Kamminga, Ed Tanner, Martin Wells, and especially Sarah Corbet.

At York University's Department of Biology I was fortunate to be surrounded by those engaged in research on symbiosis. My supervisors Peter Young and Richard Law were extraordinarily patient as I grappled with my recalcitrant microbial associates. As members of my research committee, Angela Douglas and Alastair Fitter encouraged me to continue even when a year's research work seemingly had led to a dead end. My lab mates Simon Edwards, Kaisa Haukka, and Sarah Turner not only offered me kind advice, but kept me laughing. At the University of East London, Jim Harris and Tom Hill provided valuable advice on soil biology and statistical modeling, respectively.

The book was completed while I was a consultant to ActionAid India and a guest of the National Centre for Biological Sciences, in Bangalore, India, whose researchers, especially Shona Chattarji, I thank for many fascinating discussions.

I would never have thought of historicizing the story of symbiosis if I had not encountered the brilliant transdisciplinary scholarship of Jan Sapp. It was through conversations with him that I came across some of the most remarkable events described in the book. The work of Gail Fleischaker, Chris Lyons, and Tim Ingold was also very helpful. Though I do not refer to Mark McMenamin's concept of Hypersea specifically, I am indebted to the way in which his theory elucidated the symbiotic continuity between life at sea and on land.

I am no great experimenter, and I owe most of my practical experiences to Marty Dworkin and John Breznak, who taught microbial diversity at the Summer School at Woods Hole's Marine Biological Laboratory; Oona West's symbiosis course, part of the Boston University Marine Program; and especially Lynn Margulis.

My interest in lichens and their symbiotic origin was fueled by two short courses run by the Field Studies Council, led by Peter James and Frank S. Dobson, who have also written beautifully on the subject. Grace Prendergast and Peter Roberts kindly took time to explain their work on orchids to me, while Alan Rayner's pioneering book on fungi acted as a spur to finish my own.

Early in my investigations, symbiologist David Lewis's insights made me realize that there were other ways that organisms could evolve intimate associations that did not involve inevitable conflict. Stephan Harding cheerfully entertained my ideas and spent precious hours with me slaving over a computer to see whether our intuitions could form the basis of an evolutionary model. Tim

Lenton, James Lovelock, Michel Pimbert, David Schwartzman, John Stolz, Tyler Volk, and Peter Westbroek have all, in their different ways, helped me see the significance of life's liaisons from a global and humanitarian perspective.

The old English adage that opens chapter 4 was suggested by a member of a Citizens' Jury I facilitated in Brighton, U.K. She remembered it from her childhood in the 1940s, and kindly shared it with me.

Over the years my research has been funded from various sources, including the NASA Planetary Biology Internship program, the Natural Environment Research Council, the Office of Naval Research, the Wellcome Trust, and the Marine Biological Laboratory at Woods Hole.

Much of the history of research into symbiosis is in long-forgotten tomes and obscure journals. The staff at numerous libraries have helped me track down the sources, and I am very grateful to them. These include librarians in the J. B. Morrell Library, University of York; the Genetics, Earth Sciences, Pathology, Plant Sciences, Scientific Periodicals, University, Whipple, and Balfour Libraries at the University of Cambridge; the Marine Biological Laboratory Library, Woods Hole; and the British Library, St. Pancras.

Index